# Advanced Concepts and Applications of Laser Pulses

# Advanced Concepts and Applications of Laser Pulses

Edited by **Juan Landers**

CLANRYE INTERNATIONAL

New Jersey

Published by Clanrye International,
55 Van Reypen Street,
Jersey City, NJ 07306, USA
www.clanryeinternational.com

**Advanced Concepts and Applications of Laser Pulses**
Edited by Juan Landers

International Standard Book Number: 978-1-63240-014-7 (Hardback)

# Contents

# Preface

The purpose of the book is to provide a glimpse into the dynamics and to present opinions and studies of some of the scientists engaged in the development of new ideas in the field from very different standpoints. This book will prove useful to students and researchers owing to its high content quality.

This book, written by renowned experts from across the globe, aims to elucidate the diverse concepts of laser pulses. The book explains characteristics of laser pulse creation, classification and applications. It even illustrates accomplishments made in designs, experiments and theories. The book contains examples of laser procedures in biomedical areas, and tremendously high power systems used for substance processing and water decontamination. This book will help students, managers and engineers to understand laser technology better.

At the end, I would like to appreciate all the efforts made by the authors in completing their chapters professionally. I express my deepest gratitude to all of them for contributing to this book by sharing their valuable works. A special thanks to my family and friends for their constant support in this journey.

**Editor**

# Unusual Applications

# Progress in High Average Power, Short Pulse Solid State Laser Technology for Compton X-Ray Sources

Akira Endo

Additional information is available at the end of the chapter

## 1. Introduction

Laser Compton X-ray source has been developing in more than decade as an accelerator-laser hybrid technology to realize a compact, high brightness short wavelength source. The basic principle is similar to an undulator emission, in which a high intensity laser field plays as the modulating electromagnetic field. Basic principle of the laser Compton X-ray source is explained in this chapter with recent examples of phase contrast imaging of bio samples. Single shot imaging is critical for many practical applications, and the required specification is explained as the laser pulse must exceeds some threshold parameters. It is already well studied on the optimization of the laser-Compton hard X-ray source by single shot base (John, 1998, Endo, 2001). Experimental results agreed well with theoretical predictions. Highest peak brightness is obtained in the case of counter propagation of laser pulse and electron beam bunch with minimum focusing area before nonlinear threshold (Babzien et.al, 2006: Kumita, et.al, 2008). The new short wavelength light source is well matured to demonstrate a single-shot phase contrast bio imaging in hard X-ray region (Oliva, et.al, 2010). The employed laser is a ps $CO_2$ laser of 3J pulse energy (Pogorelsky, et.al, 2006), but the laser system is not an easy and compact one for further broad applications in various laboratories and hospitals.

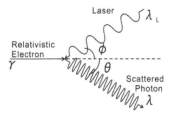

**Figure 1.** Schematic of laser-Compton scattering process

The major challenge of the laser Compton source for single shot imaging is the generation of threshold X-ray brightness, which in turn results in a clear sample imaging. Figure 1 describes the schematic of the laser-Compton interaction between electron beam and laser.

Laser-Compton scattering photon spectrum has a peak in the forward direction at a wavelength;

$$\lambda_p = \frac{\lambda_L(1 + \frac{K^2}{2})}{2\gamma^2(1 + \beta cos\phi)} \tag{1}$$

where $\gamma$ and $\beta$ are Lorentz factors, $\lambda_L$ is the laser undulation period (laser wavelength), $K$ is the K parameter of the undulator, which is equivalent to the laser intensity parameter, and $\Phi$ is the angle between electrons and laser propagation direction. The spectrum depends on the angular distribution; the wavelength $\lambda$ is emitted at

$$\theta = \frac{1}{\gamma}\sqrt{\frac{\lambda - \lambda_p}{\lambda_p}} \tag{2}$$

It is seen that higher $\gamma$ electron beam produces higher brightness of generated X-ray beams. The general formula of obtainable X-ray photon flux $N_0$ is calculated in the normal collision by the following expression,

$$N_0 \propto \frac{\sigma_c N_e N_p}{4\pi r^2} \tag{3}$$

where $\sigma_c$ is the Compton cross section ($6.7 \times 10^{-25}$ cm$^2$), $N_e$ is the total electron number, $N_p$ is the total laser photon number, and $r$ is the interaction area radius. Longer wavelength laser like ps $CO_2$ laser is advantageous to generate higher brightness X-rays at a fixed wavelength due to higher $\gamma$ factor of employed electron beam, namely higher energy accelerator. Same energy laser pulse contains 10 times photons compared to solid state laser ones. Disadvantage is that the total system size becomes larger compared to the case of solid state laser based Compton source.

The approach to increase the photon flux is equivalent to increase $N_e$, $N_p$ and decrease $r$, but there are instrumental limitations to realize these simultaneously. The practical limitation is the maximum electron number $N_e$ and minimum interaction area diameter $r$. These are determined by emittance of the accelerated electron bunch and Coulomb repulsion. We would like to suppose it as 1nC, 3ps and focusable down to 10μm diameter at 38MeV acceleration energy. Another limitation is the onset of the nonlinear threshold of the higher harmonics generation, which is evident over $10^{17}$W/cm$^2$ $CO_2$ laser irradiation intensity (Kumita, et.al. 2008). Laser pulses with 1ps pulse width focused down to 10μm, reaches at this threshold with 100mJ pulse energy. The nonlinear Compton threshold is characterized by the laser field strength

$$a_0 = eE/m\omega_L c \qquad (4)$$

where $E$ is the amplitude of laser electric field, $\omega_L$ is the laser frequency and $c$ is the speed of light. The laser field strength is linearly depending on the laser wavelength. The laser energy for the nonlinear threshold of $a_0 \sim 0.6$ corresponds to 1J with 1ps at 10µm focusing in case of solid state laser. Single shot imaging was already realized by a 3J, 5ps $CO_2$ laser pulse focused onto 0.5nC, 32µm electron bunch (Oliva, et.al 2010). The focused laser intensity is over the nonlinear threshold as $a_0 > 1$. The X-ray spectrum was evidently overlapped with higher harmonics of X-rays. We can then estimate as it is also possible to expect a single shot imaging with equivalent solid state laser pulse, once it is possible to focus down to 10µm diameter to overcome the magnitude lower laser photon number. Table 1 summarizes the design laser parameters optimized for single shot imaging. It is clear from the table that a one pulse configuration is not possible to realize a single shot imaging because of the nonlinear threshold.

| Nonlinear threshold | 1J |
|---|---|
| Single shot imaging | 4J |
| Pulse width | 1ps |
| Focus diameter | 10µm |

**Table 1.** Solid state laser parameters for single shot imaging by Compton X-ray source

Usual approach is to increase the repetition rate of the event, and the obtainable X-ray photon average flux is expressed as;

$$N = f \times N_0 \qquad (5)$$

where f is the repetition frequency. Fundamental characterization of the laser-Compton X-ray source has been undertaken with f typically as 1-10 Hz. High flux mode requires f in 100MHz range in burst mode for an equivalent single shot imaging.

The first approach is the pulsed laser storage in an optical enhancement cavity for laser-Compton X-ray sources (Sakaue, et.al. 2010, 2011). The enhancement factor P inside the optical cavity was 600 (circulating laser power was 42kW), in which the Finess was more than 2000, and the laser beam waist of 30µm ($2\sigma$) was stably achieved using a 1µm wavelength Nd:Vanadium mode-locked laser with repetition rate 357MHz, pulse width 7ps, and average power 7W. The schematic of the employed super-cavity is shown in Figure 2.

Short laser pulse *input* is injected through mirror 1 with transmittance $T_1$ and reflectance $R_1$. The mirror curvature is given as $\varrho$. The beam waist is given as $W_0$ and the cavity length is given as $L_{cav}$. The injected pulses overlap with the following pulses inside the cavity indicated as *Stored*. The loss is caused due to transmissions $T_1$ and $T_2$ of both mirrors.

An enhancement cavity requires high reflectivity and low transmittance mirror i.e. ultra-low loss mirror as an input and high reflectivity mirror as an output for high enhancement. The enhancement P is expressed by using cavity finesse F as (Hodgson, et.al, 2005);

$$P = \frac{F}{\pi} \qquad (6)$$

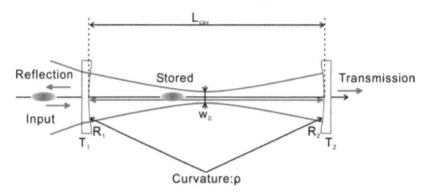

**Figure 2.** Schematic of laser storage enhancement cavity of Sakaue

It is noted that the assumed cavity length is perfectly matched with the repetition rate of input laser pulses. Finesse $F$ is given by;

$$F = \frac{\pi\sqrt{R_{eff}}}{1 - R_{eff}} \tag{7}$$

where $R_{eff}$ is $\sqrt{R_1 R_2}$. As is described above, higher reflectivity provides a higher enhancement cavity. Particularly the loss, which includes both absorption and scattering on the reflection coating, is the critical issue for storing high power laser beam. The beam waist of an enhancement cavity is described as;

$$w_0^2 = \frac{\lambda}{\pi}\sqrt{\frac{L_{cav}(2\rho - L_{cav})}{2}} \tag{8}$$

where $\lambda$ is the wavelength of the laser, $L_{cav}$ is the cavity length, $\rho$ is the curvature of the cavity mirror. While high enhancement is relatively easier, smaller waist cavity down to 10μm is difficult as described in Eq. (8). Another work reported an enhancement of $P\sim1400$ with a 22 μm beam waist and 72kW storage power (Pupeza, et.al. 2010). The scaling limit is given by optics damage, which is around 100kW with ps pulse in this research stage. It was reported by Sakaue on an imaging demonstration by using the enhancement cavity approach of Fig.2, in which the stored pulse energy was 200μJ level in a burst mode of 100 pulses. The equivalent macro pulse energy was 20mJ. The larger focusing spot decreased available X-ray photons in each collision event, and the required time for imaging was much longer than equivalent single shot imaging (Sakaue, et.al, 2012). The repetition rate was 3Hz, and imaging of a fish bone was taken in 30 min with total laser energy of 108J. Figure 3 shows an imaging example in this experiment. Once the laser is focused to 10 μm diameter, and electron beam is focused to 30μm diameter, then the required total laser energy deceases to 3J, which indicates the design parameter of Table 1 as a good measure.

Grating based X-ray phase contrast imaging is now developing as a more sensitive imaging technology (Momose, et al. 2012), and a high repetition rate X-ray source, based on an enhancement cavity combined with a compact synchrotron, was recently introduced in preclinical demonstration of biological samples (M.Bech, et.al 2009). The X-ray peak energy was 13.5keV with 3% band width. The source size was relatively large as 165µm due to the focusing limit of circulating electron bunch in the compact ring. The repetition rate was typically in continuous 100MHz region, but the unit imaging time period was around 100 seconds (~2 minutes) due to lower X-ray photon flux per each event.

Classical low repetition rate laser Compton X-ray source demonstrated earlier a successful in-line phase contrast imaging of biological samples (Ikekura-Sekiguchi,et.al 2008). The repetition rate was at 10Hz with 40µm diameter source size. The imaging was undertaken by 3ps pulse width X-ray beam of 30keV energy. The required shot number for imaging was 18000 (30 minutes). It was indicated by this experiment that a solid state laser must have higher pulse energy more than 1J, and a better beam quality for 10µm focusing, for single shot imaging. We evaluate a possible solid state laser technology in the following sections on this subject, by reviewing practical instrumental limitations and propose the most promising approach for a compact single shot laser-Compton X-ray imaging.

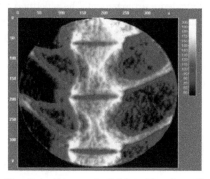

**Figure 3.** Refraction contrast imaging of bio sample (fish bone) by a laser-Compton X-ray source (Sakaue, et.al. 2012)

## 2. Temporal and spatial synchronization between electron beam and laser pulses

The essential technology for the laser-Compton X-ray source has been well studied in the Femtosecond Technology Project in Japan, and the achieved performance of the X-ray beam was also well characterized. Mathematical formula was obtained on its fluctuation depending on the temporal and spatial jitters (Yorozu, et.al 2002). Synchronization and stabilization technology was developed to the stage that the resulting pulse–pulse X-ray fluctuation almost reflects the laser pulse energy fluctuation (Yanagida, et.al 2003). The achieved overall performance was reported by T.Yanagida in a SPIE conference (Yanagida, et.al 2005). Figure 4 and table 2 show the system configuration and the summary of the specification of the laser-Compton X-ray source, studied and developed in the FESTA program. A phase contract imaging was also demonstrated by this light source of bubbles in solidified adhesives.

The electron beam is generated from a photo cathode RF gun driven by a synchronized picosecond UV laser, and accelerated to 38MeV energy by a S-band Linac. The achieved normalized emittance was 3 $\pi$mm-mrad, and resulted in the focused beam size as 30$\mu$m. It was demonstrated as further reduction of emittance was possible by spatial and temporal shaping of irradiation laser pulse for electron beam from photo cathode (Yang, et.al.2002). The employed laser for X-ray generation was a 4TW Ti:Sapphire laser with 800nm wavelength. The laser pulse was focused down to 10$\mu$m diameter and the peak intensity was around $10^{18}$ w/cm$^2$. The number of generated X-rays was measured with Micro Channel Plate located 2.6m downstream from the interaction point (source point). The MCP gain was calibrated using a standard $^{55}$F X-ray source with known strength. The pulse width was estimated from measured electron beam and laser pulse width. The X-ray pulse width is almost determined by longer electron beam pulse width in case of normal incidence (165°interaction angle) and the cross section of the focused electron beam in case of 90°interaction angle. The long term fluctuation of the generated X-ray pulses is shown in Figure 5 in case of normal incidence arrangement. The repetition rate was 10Hz and the X-ray fluctuation was 6%, which is almost equivalent to the fluctuation of incident laser pulse energy. The laser focused intensity is around the nonlinear laser-Compton threshold as a0~0.6. This was confirmed by a calculation by CAIN code in Figure 6. It is observed in the calculation of a nonlinear effect in the higher component of the generated X-ray energy distribution by blue dots (calculation by K.Sakaue).

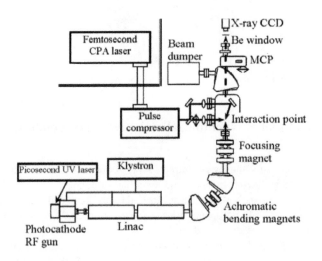

**Figure 4.** System configuration of laser-Compton X-ray source

| Electron parameters | | X-ray parameters | | |
|---|---|---|---|---|
| Electron energy | 38 MeV | Interaction angle | 165° | 90° |
| Bunch charge | 0.8 nC | Maximum energy | 33.7 keV | 17.1 keV |
| Bunch width | 3 ps (rms) | Total number of photons | $2\times10^6$ | $5\times10^5$ |
| Beam size | 30 µm (rms) | (photons/pulse) | | |
| Normalized emittance | 3 π mm·mrad | Pulse width | 3 ps (rms) | 150 fs (rms) |
| | | Intensity fluctuation | 6% | 11% |
| Laser parameters | | Repetition rate | 10 Hz | 10 Hz |
| Pulse energy | 200 mJ | | | |
| Pulse width | 50 fs (FWHM) | | | |
| Wavelength | 800 nm | | | |
| Beam size | 10 µm (rms) | | | |

**Table 2.** Summary of the electron beam and laser parameters and obtained X-ray parameters

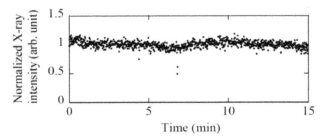

**Figure 5.** Fluctuation of X-ray intensity in the normal incidence laser-Compton X-ray generation

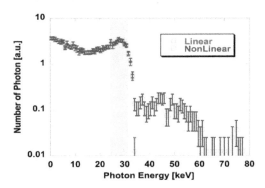

**Figure 6.** Calculated results of the X-ray energy distribution by CAIN code in log plot for linear (red) and nonlinear (blue) laser Compton scattering. Higher energy components are accompanying the linear ones. Light green region indicates the x-ray spectrum employed for imaging.

It is noticed that the component technologies for a single shot imaging by laser-Compton X-ray is well matured. There are but still several concerns necessary to design an optimized multi pulse method to realize the threshold (effective) laser energy of 4J in 10µm focus spot overlapped with electron bunch. The spatial stability of the laser-Compton X-ray source is essentially guaranteed in the order of the focus spot, because laser and electron beam must synchronize spatially (also temporally) each other to generate X-ray beam. Stable multi pulse electron beam generation is needed for efficient and stable laser-Compton X-ray source, to avoid higher harmonics noise of X-rays by limiting laser pulse intensity in each interaction. The RF photocathode gun is irradiated by synchronized ps laser pulses to generate flat top electron beam pulse train. An earlier experiment was reported by T.Nakajyo in 2003 of 60 micro pulses generation with a flat-top shape (Nakajyo, et.al 2003). The essential technology is temporal modulation of the seed laser pulse trains by Pockels Cell, to compensate the amplification saturation of the seeded pulse trains in the power amplifiers. Figure 7 shows the example of the pulse train amplification without and with intensity modulation. The obtained flatness of the 60 bunch electron beam was equal to that of the incident laser train (<7%) during 0.5μsec duration. The time duration is regarded for bio imaging enough short for effective single shot imaging.

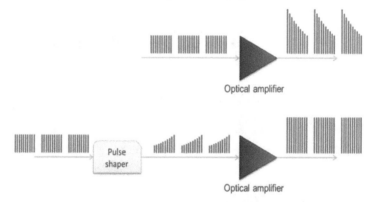

**Figure 7.** Pulse train amplification without (upper) and with (lower) intensity modulation

The other consideration is the selection of the amplifier module. It is required to focus 1J, 1ps laser pulses onto 10μm spot in spatial multiplexing in a near normal incidence arrangement. The requirement for the beam quality is expressed by $M^2$ parameter of the laser beam. Figure 8 shows the relationship between $M^2$ and focused beam spot size with the beam diameter as the working parameter. It is clear that $M^2$ is required to be less than 2 to realize a 10μm focal spot diameter with 20mm original beam diameter.

## 3. Thermal distortion in solid state amplifier

The basic requirement for a laser driver in a single shot laser-Compton X-ray imaging is summarized as Table 3. The $M^2$ of the laser beam is less than 2 from the discussions in the section 2. Detailed design work is required on the spatial configuration of 8 focusing optics to satisfy the 10 $\mu$ m focus spot by avoiding radiation damage to the optics from scattered X-ray and electrons. The main consideration is an evaluation of the innovative solid state laser technology in the last decade like fiber, thin slab and thin disc lasers.

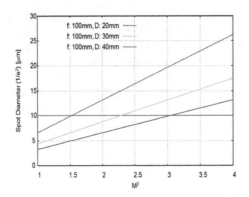

**Figure 8.** $M^2$ and focused beam size with beam diameter as parameters

| Module pulse energy | >500mJ/ps |
|---|---|
| Module number | 8 units |
| Multiplexed energy | 4 J |
| Micro pulse time interval | 8.4ns (119MHz) |
| Macro pulse width | ~60ns |

**Table 3.** Laser parameters for a laser driver in a single shot laser-Compton X-ray imaging around 30keV

Laser diode pumped rod type laser was regarded as the most suitable laser to meet the simultaneous requirement of high pulse energy, high average power together with high beam quality, before the fundamental solid state laser innovation. It was well known that flush lamp pumped solid state laser suffered from high thermal distortion of the laser medium due to low optical-optical conversion efficiency. Laser diode pumping was expected to solve the thermal distortion problem by improved energy conversion efficiency in the same configuration. Figure 9 is an example of a LD pumped rod Nd:YAG amplifier of 9mm diameter. Maximum LD pump power was 2.1kW, and optical-optical conversion efficiency was 41% (Endo et.al, 2004).

The fundamental difficulty of the LD pumped large rod amplifier comes from slow cooling speed of the laser material from the water jacked located around the rod. The resulting temperature gradient causes thermal lensing, which is expressed analytically by the following expression (Koechner, 1999).

$$f = \frac{KA}{P_a}\left(\frac{1}{2}\frac{dn}{dT} + aC_{r,\phi}n_0^3 + \frac{ar_0(n_0-1)}{L}\right)^{-1} \tag{9}$$

Temperature profile becomes radially parabolic. The first term corresponds to the temperature depending refraction index change of 70% contribution to $f$, the second term is stress induced refraction index change of 20% contribution to $f$, and the last term is temperature depending surface effect of 10% contribution. The cumulative effect of beam amplification and propagation of spatially non uniform beam results, combined with slight non uniform initial gain distribution, in a chaotic wave front with higher $M^2$. Figure 10 shows an example of beam cross section after booster amplifier of Fig.9 with 1.1kW average power at 10 kHz repetitive amplification of 6ns pulses. The beam was focused by f=10cm lens to 350 µm diameter with 10mm initial beam diameter. The resulting $M^2$ was nearly 35. The fundamental problem of rod amplifiers comes from temperature gradient. This was the main motivation of the enthusiastic search for a new architecture of low $dn/dT$ solid state laser technology in the last two decades (Injeyan. et.al, 2011).

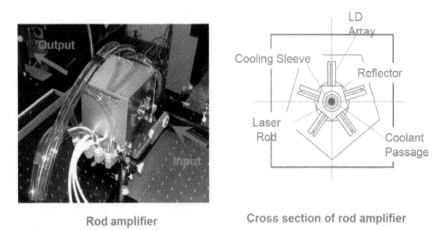

**Figure 9.** Outlook and cross section configuration of 9mm diameter Nd:YAG rod amplifier

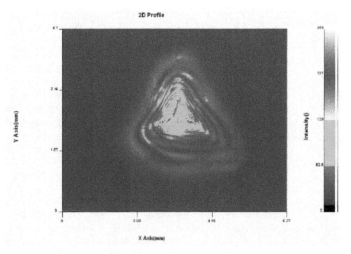

**Figure 10.** Beam shape after rod amplifier of 9mm diameter at 1.1kW average power

## 4. Thin disc laser as a high beam quality, short pulse solid state amplifier

Cryogenic cooling was considered to solve the temperature gradient problem in rod type LD pumped laser. MIT laser scientists are working following on this concept with recent unprecedented results of $M^2$<1.05 from cryogenically cooled (77k) bulk Yb:YAG laser in Q-switched mode of 20mJ/16ns at 5kHz. The average power was modest 100W in this experiment. The pointing stability was reported as 20 $\mu$ radian as mean deviation (Manni, et.al, 2010). One disadvantage of the cryogenic cooled Yb:YAG is the gain bandwidth narrowing, and compression to 1ps pulse width is not appropriate due to this effect (Hong, et.al. 2008). Fiber laser technology is progressing significantly with various laser specifications in CW and pulsed mode due to its efficient cooling characteristics owning to larger surface area/volume ratio. One drawback of fiber laser is its limited short pulse energy due to smaller medium diameter. There are still significant progresses in this field from its early work by a cladding-pumped, Yb doped large core fiber amplifier with specifications of 50W average power by 80MHz repetition rate of 10ps pulses with $M^2$<1.3 (Limpert, et.al,2001). Recent experiments achieved high pulse energy of 26mJ with 60ns pulse width at 5kHz repetition rate in Q-switched mode by a large-pitch fiber with a core diameter of 135 $\mu$ m (Stutzki,et.al. 2012). The average power is approaching to kW level with femtosecond pulse at high repetition rate. The reported performance was $M^2$=1.3 with 0.9ps pulse width with average power 830W at 80MHz repetition rate (Limpert, et.al. 2011). Sandwiched thin slab geometry is also promising to realize low temperature gradient inside laser medium by efficient cooling from both sides of thin slab. Multi-pass amplification is successfully employed inside the medium with expanding beam shape to keep the laser intensity constant during amplification. 1.1kW average power was reported with Yb:YAG as laser medium. The repetition rate was 80MHz of 615fs pulses, with $M_x$=1.43 and $M_y$=1.35 (Russbueldt, et.al. 2010). All these approaches are remarkable, especially regarding the beam quality $M^2$, but the achievable pulse width, and energy is limited due to cryogenic temperature or limited beam diameter in each technology. It is to be noticed that kW level, 80MHz femtosecond source could improve the average stored laser power in an enhancement cavity, once present limitation of optics damage is eliminated.

Thin disc laser is characterized with its larger diameter, and fundamentally suited for high pulse energy amplification. The schematic of a thin disc laser is shown in Fig.11. Thin disc of laser active medium like Yb:YAG of typical diameter 25mm is molded on a high reflectivity mirror (both wavelength of multi-pass LD; 940nm and laser wavelength; 1030nm). Water cooling from the backside of thin disc keeps the medium temperature around 15 degree. Mechanical distortion of the surface and ASE gain depletion is the main subject to be considered for high beam quality, short pulse high energy amplification. There are several activities to realize one J pulse energies with $M^2$<1.3 in ps pulse length at high repetition rate by thin disc laser technology. Conceptual design of a spatially multiplexed laser driver for single shot laser Compton imaging is presented in the next section, by showing several research examples.

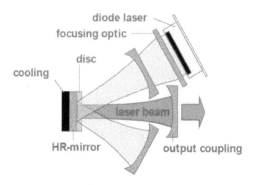

**Figure 11.** Schematic of thin disc laser configuration

## 5. Laser driver for single shot laser-Compton imaging

Candidate materials are considered for this particular application as Yb:S-FAP (Yb:Sr5(PO4)3F) or Yb:YAG. Comparison of both material characteristics are shown in table 4 (Payne,et.al, 1994). Both are characterized with higher quantum efficiency (Stokes factor), which is advantageous to less thermal stress after pulse energy depletion. Crystal growth to a larger diameter is important to avoid laser induced damage on the laser medium surface for 1J, ps pulse amplification at high repetition rate. It was tried to select Yb:S-FAP as the laser material by an end pumped square bar configuration, for the development as the future laser driver for high brightness laser Compton X-ray source (Ito et.al, 2006). The oscillator was a Yb:glass mode locked laser with 200fs, 170mW average power at 79.33MHz repetition rate, tuned at 1043nm wavelength. The oscillator pulse was stretched by a grating pair, and seeded into a cavity of a regenerative Yb:S-FAP laser by a Pockels Cell. Stacked laser diode array irradiated the Yb:S-FAP square rod (3.5 x 3.5 x 21 mm$^3$) with 900 nm wavelength, for 1.3ms duration of 1J pulse energy, through a lens duct and aspheric lens. The regenerative amplifier delivered 24mJ and the pulse was compressed down to 2ps in an initial experiment. Pre-amplifiers and main amplifiers were designed on the same architecture. Main amplifier employed square rods of geometrical size as 8 x 8 x 24 mm$^3$. Heat removal at higher repetition rate was not efficient from these amplifiers and the amplification was not perfect due to thermally induced birefringence. It was recently reported that "Mercury Laser Program" has achieved 100J in ns pulse length at 10Hz repetition rate from a side pumped thin slab Yb:S-FAP module of 3cm x 5cm aperture with a powerful cooling by He gas flow (Ebbers, et.al 2009). It is essentially proved from these experiments that Yb:S-FAP is usable as a laser material for specific ps application with higher pulse energy, once a large gas flow system is allowed in the whole system.

Another candidate is Yb:YAG for short pulse, high repetition rate operation for various applications. It is discussed that there is an obstacle to obtain large pulse energy in J level, from a bulk structure Yb:YAG material like a rod due to thermal population of the lower laser level (Ostermeyer, et.al. 2007). Solution might be found in a new configuration

|  | Yb:S-FAP | Yb:YAG |
|---|---|---|
| **Pump wavelength (nm)** | 900 | 940 |
| **Laser wavelength (nm)** | 1047 | 1030 |
| **Fluorescence lifetime (ms)** | 1.26 | 1.0 |
| **Emission cross section ($10^{-20}$cm²)** | 7.3 | 2.3 |
| **Saturation fluence (J/cm²)** | 3.2 | 9.6 |
| **Pump saturation intensity (kW/cm²)** | 2.3 | 32 |
| **Spectral bandwidth (nm)** | 3.5 | 9.5 |
| **Thermal conductivity (W/mK)** | 2 | 10 |

**Table 4.** Specific characteristics of Yb:S-FAP and Yb:YAG materilas

optimized for efficient cooling. Thin disc configuration is advantageous for the sake of efficient heat removal from gain media. It was tried to develop a pulsed thin disc laser with 1kW average power at 10 kHz repetition rate (Miura, et.al. 2005). Cavity optimization was performed for a regenerative amplifier, composed of two Yb:YAG thin disc modules, by compensating the deformation of optical components inside the cavity, with high beam quality at 500W CW operation. The extinction rate of linear polarization was more than 1:140. The developed regenerative amplifier module was connected with a seeder, which was a Yb:glass mode locked oscillator with 325fs pulse width and a fiber pulse stretcher. The extended pulse was injected into the regenerative amplifier cavity at 10 kHz repetition rate. The experimental configuration is shown in Fig12. Figure 13 is the pulse build up inside the regenerative amplifier cavity. Output average power was 33W in single mode, and 73W in multi mode with 50-100 ps pulse length (Miura,et.al. 2006). It was reported that an average power of 75W was achieved at 3kHz repetition rate with pulse energies exceeding 25mJ, a pulse-pulse stability of <0.7% (rms), a pulse duration of 1.6ps from an improved single thin disc module configurated in a regenerative amplifier with high beam quality as $M^2$<1.1 (Metzger, et.al 2009).

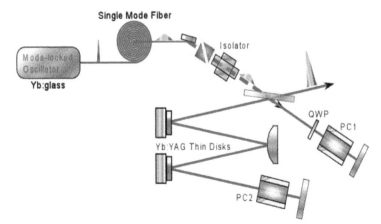

**Figure 12.** Schematic of a dual module thin-disc regenerative amplifier

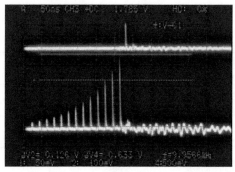

**Figure 13.** Operation of thin disc regenerative amplifier. Upper trace shows sliced out pulse, and lower trace shows building up inside cavity

Pulse energy increase to J level needs a multi pass amplifier without intra cavity Pockels Cell. The thickness of a thin disc medium is less than mm length for efficient water cooling from back side, and the single pass gain is lower than that of a rod medium in general. Multi pass optical cavity is required for this purpose, without any beam distortion during the amplification. A study was tried to design an optimized multi pass mechanical structure (Neuhaus, et.al.2008). A progress was recently reported from a group of Max Born Institute, Berlin, Germany on a development of a diode pumped chirped pulse amplification (CPA) laser system based on Yb:YAG thin disk technology, with a repetition rate of 100 Hz and output pulse energy aiming in the joule range (Tuemmler, et.al, 2009). Regenerative amplifier pulse energy was more than 165 mJ at a repetition rate of 100 Hz with a stability of 0.8% over a period of more than 45 min. The optical to optical conversion efficiency was 14%. The following main amplifier increased pulse energy to more than 300 mJ by a multi pass configuration. A nearly bandwidth limited recompression to less than 2 ps was also demonstrated. Further scaling of this technology is possible by enlargement of the thin disc diameter by careful optimization of the mitigation of surface deformation and ASE gain depletion. The latter phenomenon is well known in a small aspect ratio laser medium (Lowental, 1986). Numerical modeling of ASE gain depletion is useful to optimize working parameters, and HiLASE project is engaged in this effort to achieve 1J level picosecond pulses with high beam quality from thin disc amplifiers (Smrz, et.al.2012).

It is possible to design a spatial-temporal multiplexing of 0.5J, 1ps pulses onto the interaction point with low emittance electron bunch as is shown in Fig.14. Multiplexing of 8 pulses in polarisation combined 4 beams is the natural configuration. Timing jitter is possible in fs range which causes no actual X-ray output fluctuation. Spatial overlapping on 10μm diameter spot is challenging with pointing stability in the 10μrad range. Figure 14 indicates the multiplexing scheme to realize the laser specification of Table 2, based on 0.5J, ps thin disc laser modules of 8 units.

The generated forward directed X-ray beam has an effective pulse width <70ns, which is enough short for single shot imaging of bio samples. It is noted that the relative interaction angle between electron bunch and laser beams are fixed as 165 degree each other, in axial symmetry. It is proposed in a white book published by ELI Nuclear Physics working group, as the first stage of gamma ray program based on laser-Compton scheme, assumes 20 micro pulses with 0.15J, ps laser pulses, which is 3J effectively (Barty,C. et.al. 2011). The macro pulse repetition rate is expected as 120Hz. The average laser power is 360W. This is a manageable specification by usable laser technology described in this article.

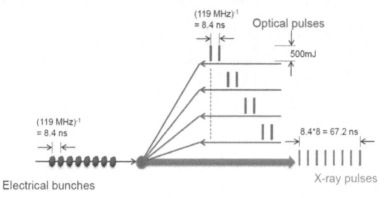

**Figure 14.** Spatial-temporal multiplexing of 0.5J,ps laser pulses

## 6. Conclusion

This chapter described the laser-Compton X-ray generator. The compact, high brightness X-ray source has been designed, fabricated and tested. This technology provides successful single-shot imaging of bio samples with multi J solid state laser pulses of ps pulsewidth. Advanced laser technologies were evaluated to realize a high beam quality, 1J level pulses. Thin disc laser was shown to be the best candidate for this application with Yb:YAG as the active medium. Spatial-temporal laser multiplexing was proposed to avoid nonlinear Compton effect. Some further research effort may bring us the realization of this technology.

The author deeply appreciates to his former colleagues in the Femtosecond Technology Project (FESTA), Extreme Ultraviolet Lithography Project (EUVA), supported by New Energy and Industrial Technology Development Organization (NEDO) in Japan to their productive and advanced works. Dr.Sakaue of Waseda University in Tokyo, Japan and Dr.Miura of HiLASE Project in Prague, Czech Republic, are especially appreciated to prepare various materials in this article.

This work was partially supported by the Czech Republic's Ministry of Education, Youth and Sports to the HiLASE project (reg.No. CZ.1.05/2.1.00/01.0027).

## Author details

Akira Endo

*Research Institute for Science and Engineering, Waseda University, Tokyo, Japan*
*HiLASE Project, Institute of Physics AS CR, Prague, Czech Republic*

# 7. References

Babzien, M. Ben-Zvi, I. Kusche, K. Pavlishin, I.V. Pogorelsky, I.V. Siddons, D.P. Yakimenko, V. (2006); "Observation of the second harmonic in Thomson scattering from relativistic electrons", Phys. Rev.Lett. 96 (2006) 054802

Barty, C. Wormser, G. Hajima, R. (2011); "Infrastructure producing high intensity gamma-rays for ELI nuclear physics Bucharest-Magurele, Romania", *The White Book of ELI Nuclear Physics Bucharest-Magurele, Romania*

Bech, M. Bunk, O. David, C. Ruth, R. Rifkin, J. Loewen, R. Feidenhaus, R. Pfeiffer, F. (2009); "Hard X-ray phase contrast imaging with the Compact Light Source based on inverse Compton X-rays", J.Synchrotron Rad. 16, pp43-47

Ebbers, C. Caird, J. Moses, E. (2009); "High-Power Solid State Lasers: The Mercury laser moves toward practical laser fusion", *www.laserfocusworld.com/article/355406*

Endo, A. Yang, J. Okada, Y Yanagida, T. Yorozu, M. Sakai, F. (2001); "Characterization of the monochromatic laser Compton X-ray beam with picosecond and femtosecond pulsewidths", *Proceedings SPIE 4502, pp100-108*

Endo, A (2004); „Performance of a 10 kHz laser produced plasma light source for EUV lithography", *Proceedings SPIE 5374, pp160-167*

Hodgson, N & Weber, H (2005). *Laser resonators and beam propagation: Fundamentals, Advanced Concepts and Applications 2nd edition,* ISBN-10: 0387400788, Springer, Berlin

Hong, K-H. Siddiqui, A. Moses, J. Gopinath, J. Hybl, J. Ilday, F.O. Fan, T.Y. Kaerntner, F.X. (2008); "Generation of 287W, 5.5ps pulses at 78MHz repetition rate from a cryogenically cooled Yb:YAG amplifier seeded by a fiber chirped-pulse amplification system", Opt.Lett. 33, pp2473-2474

Ikekura-Sekiguchi, H. Kuroda, R. Yasumoto, M. Toyokawa, H. Koike, M. (2008); "In-line phase contrast imaging of a biological specimen using a compact laser-Compton scattering-based x-ray source", Appl.Phys.Lett. 92, 131107(2008)

Injeyan, H. Goodno, G (2011); *High power laser handbook,* ISBN-978-0-07-160901-2, McGrow-Hill, New York

Ito, S. Nakajyo, T. Yanagida, T. Sakai, F. Endo, A. Torizuka, K. (2006); "Diode-pumped chirped pulse Yb:S-FAP regenerative amplifier for laser-Compton X-ray generation" Opt.Commun. 259, pp812-815

John, R.W (1998); "Brilliance of X rays and gamma rays produced by Compton back scattering of laser light from high energy-electrons", Laser and Particle Beams, 16 (1998) 115-127

Koechner, W. (1999); *Solid-State Laser Engineering,* Springer, Berlin

Kumita, T Kamiya, Y Babzien, M Ben-Zvi, I Kusche, K Pavlishin, I.V. Pogorelsky, I.V. Siddons, D.P. Yakimenko, V. Hirose, T Omori, T Urakawa, J Yokoya, K Cline, D. and Zhou, F (2008); "Observation of the Nonlinear Effect in Relativistic Thomson Scattering of Electron and Laser Beams", Laser Phys. 16 pp267-271

Lowenthal, D.D. Egglestone, J.M. (1986); "ASE effects in small aspect ratio laser oscillators and amplifiers with nonsaturable absorption", IEEE J.Quantum Electron. QE-22, pp1165-1173

Limpert, J. Liem, A. Gabler, T. Zellmer, H. Tuennermann, A. Unger, S. Jetschke, S. Mueller, H.R. (2001); "High average power picosecond Yb-doped fiber amplifier", Opt.Lett.26, pp1849-1851

Limpert, J. Haedrich, S. Rothhardt, J. Krebs, M. Eidon, T. Schreiber, T. Tuennermann, T. (2011) " Ultrafast fiber lasers for strong-field physics experiments" Laser Photonics Rev. 5 pp634-646, DOI 10.1002/lpor.201000041

Manni, J.G. Hybl, J.D. Rand, D. Rippin, D.J. Ochoa, J.R. Fan, T.Y. (2010); "100-W Q-switched cryogenically cooled Yb:YAG laser", IEEE J.Quantum Electron. QE-46, pp95-98

Metzger, T. Schwarz, A. Teisset, C.Y. Sutter, D. Killi, A. Kienberger, R. Krausz, F. (2009); "High-repetition-rate picosecond pump laser based on a Yb:YAG disc amplifier for optical parametric amplification, Opt.Lett. 34, pp2123-2125

Miura, T. Suganuma, T. Endo, A. (2005) ; "High average power pulsed Yb:YAG thin-disc laser", *Proceedings SPIE, 5707, pp91-98*

Miura, T. Endo, A. (2006); "High power Yb:YAG thin-disc laser for EUV lithography", *Proceedings of 2nd EPS-QEOD 2006 Europhoton Conference, ThD5, 10-15 September, 2006, Pisa, Italy*

Momose, A (2012); International workshop on X-ray and neutron phase imaging with gratings, March 5-7, 2012 Tokyo, Japan

Nakajyo, T. Yang, J. Okada, Y. Yanagida, T. Yorozu, M. Sakai, F. Aoki, Y. (2003); "Multi-bunch electron beam source with a magnesium-photocathode radio-frequency gun", *Proceedings of RadTech ASIA 2003, pp.234-237 (December 9-12, 2003, Yokohama, Japan)*

Neuhaus, J. Kleinbauer, J. Killi, A. Weiler, S. Sutter, D. Decorsy, T. (2008) ; "Passively mode-locked Yb:YAG thin-disc laser with pulse energies exceeding 13µJ by use of an active multipass geometry", Opt.Lett. 33, pp726-728

Oliva, P. Carpinelli, M. Golosio, B. Delogu, P. Endrizzi, M. Park, J. Pogorelsky, I. Yakimenko, V. Williams, O. Rosenzweig, J (2010); "Quantitative evaluation of single-shot inline phase contrast imaging using an inverse Compton x-ray source", Appl.Phys.Lett. 97, 134104

Ostermeyer, M. Straesser, A. (2007); "Theoretical investigation of feasibility of Yb:YAG as laser material for nanosecond pulse emission with large energies in the joule range", Opt.Commun. 274, pp422-428

Payne, S.A. Smith, L.K. Deloach, L.D. Kway, W.L. Tassano, J.B. Krupke, W.F. (1994); "Laser, optical and thermomechanical properties of Yb-doped fluorapatite", IEEE J. Quantum Electron. QE30, pp170-179

Pogorelsky, I.V. Babzien, M. Pavlishin, I. Stolyarov, P. Yakimenko, V. Shkolnikov, P. Pukhov, A. Zhidkov, A. Platonenko, V.T. (2006); "Terwatt $CO_2$ laser; a new tool for strong field research", *Proceedings of SPIE, vol 6261, 626118*

Pupeza, I. Eidam, T. Bernhardt, B. Ozawa, A. Raushenberger, J. Fill, E. Apolonski, A. Udem, T. Limpert, J. Alahmed, Z.A. Azzeer, A.M. Tuennermann, A. Haensch, T.W. Krausz, F. (2010); "Power scaling of a high repetition rate enhancement cavity", Opt.Lett. 35, pp2052-2054

Sakaue, K. Washio, M. Araki, S. Fukuda, M. Higashi, Y. Honda, Y. Omori, T. Taniguchi, T. Terunuma, N. Urakawa, J. Sasao, N (2009); "Observation of pulsed x-ray trains produced by laser-electron Compton scatterings", Rev.Sci.Instrum. 80 123304 1-7

Sakaue, K. Araki, S. Fukuda, M. Higashi, Y. Honda, Y. Sasao, N. Shimizu, H. Taniguchi, T. Urakawa, J. Washio, M. (2011); "Development of a laser pulse storage technique in an optical super-cavity for a compact X-ray source based on laser-Compton scattering", Nucl.Instrum.Meth. A637 S107-S111

Sakaue, K. Aoki, T. Araki, S. Fukuda, M. Honda, Y. Terunuma, N. Urakawa, J. Washio, M (2012);"First refraction contrast imaging via laser-Compton scattering X-ray at KEK", *Proceedings of International workshop on X-ray and neutron imaging with gratings, P36, March 5-7, 2012, Tokyo, Japan*

Smrz, M. Severova, P. Mocek, T (2012); "Design and modelong of kW-class thin disc lasers", *Proceedings to be published, SPIE Photonics West, Jan 2012, San Francisco*

Stutzki, F. Jansen, F. Liem, A. Jauregui, C. Limpert, J. Tunnermann, A. (2012) "26mJ, 130W Q-switched fiber-laser system with near-diffraction-limited beam quality", Opt.Lett. 37, pp1073-1075

Tuemmler, J. Jung, R. Stiel, H. Nickles, P.V. Sandner, W. (2009); "High-repetition rate chirped-pulse-amplification thin disc laser system with joule level pulse energy", Opt.Lett. 34, pp1378-1380

Yang, J. Sakai, F. Yanagida, T. Yorozu, M. Okada, Y. Takasago, K. Endo, A. Yada, A. Washio, M. (2002); "Low emittance electron beam generation with laser pulse shaping in photocathode radio-frequency gun", J.App.Phys. 92, pp1608-1612

Yanagida, T. Kobayashi, Y. Maeda, K. Ito, S. Sakai, F. Torizuka, K. Endo, A.(2003); "Synchronization of two different repetition rate mode-locked laser oscillation for Laser-Compton X-ray generation", *Proceeding of SPIE, 5914 pp149-156*

Yanagida, T. Nakajyo, T. Ito, S. Sakai, F (2005); "Development of high brightness hard X-ray source by Laser-Compton scattering", *Proceedings of SPIE 5918, 59180V*

Yorozu, M. Yang, J. Okada, Y. Yanagida, T. Sakai, F.Takasago, K. Ito, S and Endo, A. (2002); "Fluctuation of femtosecond X-ray pulses generated by a laser-Compton scheme", Appl.Phys.B74, pp327-331

Yorozu, M Yang, J Okada, Y Yanagida, T Sakai, F Ito, S and Endo, A (2003); "Spatial beam profile of the femtosecond X-ray pulses generated by a laser-Compton scheme", Appl.Phys. B76 pp293-297

# Chapter 2

# Ion Acceleration by High Intensity Short Pulse Lasers

Emmanuel d'Humières

Additional information is available at the end of the chapter

## 1. Introduction

### 1.1. Context

Lasers can now deliver short and very intense pulses capable of producing hot and dense plasmas. The diffusion of such lasers has allowed studying the behaviour of ionized media in novel ways. Ion acceleration through laser plasma interaction is a recent research subject and its fast-paced progresses highlight the mastering of high power laser systems and the better understanding of the physics of high intensity laser plasma interaction. Short and intense laser pulses have also permitted the development of areas of research like harmonic generation, hard X-ray radiation sources, energetic particles sources and inertial confinement fusion. These researches have lead to numerous applications in transdisciplinary domains: material science, development of X-UV lasers, surface treatments, chemistry and biology are just a few examples. Large national programs are focused on these topics like the National Ignition Facility in California, the Laser Mégajoule and the Apollon laser in France.

These topics use plasmas where kinetic and collective effects are dominant with numerous non linear phenomena. Impressive progresses are being accomplished in the modelling of these phenomena and in their optimization thanks to Particle-In-Cell (PIC) codes. This is due to the fact that the mechanisms at play are complex and difficult to model analytically, or using a fluid model. PIC codes used to describe plasmas solve the coupled system of Maxwell's equations (for electromagnetic fields) and Vlasov equations (for each particle specie). The PIC method consists in solving Vlasov equation for each particle specie by the characteristics method for macro-particles, each representing a given amount of real particles but that conserves the mass and the charge of the particles of the specie that is studied. Several research areas of high intensity laser plasma interaction require PIC codes capable of modelling atomic physics processes (ionization and collisions for instance) and energy loss processes through radiation of energetic charged particles in intense electromagnetic fields.

## 1.2. Ion acceleration in laser-plasma interaction

Particle accelerators play an important role in numerous scientific programs. Thanks to their versatility, they have found numerous applications in science and medical fields. Plasmas can endure very large acceleration gradients, making them very interesting as accelerating media. Particle acceleration by interaction of a high intensity laser with a gas or a solid is therefore attracting a growing interest. More and more laboratories in the world possess their own laser system and the results obtained by these laboratories, both theoretical and experimental, are very promising. It is nowadays possible to accelerate protons at energies of several tens of MeV [Snavely et al., 2000], and electrons at energies higher than a GeV [Leemans et al., 2006]. Moreover, the observation of intense energetic ion beams [M. Borghesi et al., 2006] represents one of the most interesting recent developments in "High Field Science", made possible by the advent of ultra-short and ultra-intense lasers [Mourou et al., 2006].

High power laser pulses open the possibility of ion acceleration to several tens MeV for protons, and higher than a GeV for Carbon ions on millimeter scale distances. These beams therefore open original perspectives, as new exploration tools of fundamental physics, or as new ways to increase the discovery potential of existing accelerators. Possible applications of these beams are numerous: ion fast ignition for inertial confinement fusion (in which a "spark", a beam of energetic ions, is used to ignite a target and reduces the total energy cost), high resolution radiography of plasmas, hadrontherapy, radio-isotopes production, laboratory astrophysics, etc... Indeed, the properties, not explored until now, in terms of current, duration (of the order of a few ps), laminarity (low divergence) and compactness – the acceleration process takes place on a distance of the order of ten microns – show that these beams are complementary, and even superior on some aspects to conventional accelerators. Finally, the constant development of high intensity lasers, on the duration, energy, repetition rates, and compactness aspects, makes the applicability perspectives of these intense sources extremely promising.

The maximum energy, as well as the properties of these ion/proton beams go well beyond the limits explored during the first experiments performed at the end of the seventies that observed ion beams from laser-matter interaction. In these first experiments, targets were irradiated with $CO_2$ ($\lambda$=10,6 $\mu$m) lasers with pulse durations < 1 ns and peak intensities $I\lambda^2$ of the order of $10^{17}$-$10^{18}$ W.cm$^{-2}$.$\mu$m$^{-2}$ [Begay et al., 1982]. The ions were emitted from the surface of the target irradiated by the laser, with a large angular divergence and an energy of a few 100 keV/nucleon. The protons originated from contaminants on the target surface. These experiments, repeated using Nd:Glass lasers at $\lambda$=1 $\mu$m produced similar results, with a higher ion energy due to the increased $I\lambda^2$ [Fews et al., 1994; Clark et al., 2000b] . Simulations using the parameters of these experiments showed that the ion acceleration process could in this case be described as a plasma expansion in vacuum mechanism, first considered isothermal [Gitomer et al., 1986; Sack et al., 1987], or adiabatic.

The discovery of energetic and collimated ion beams was made thanks to a recent fortuitous observation. The related experimental results were published in 2000 and obtained on the

LLNL PetaWatt system in the USA. The scientists conducting the experiments wanted to study the electron beams generated during the interaction of this laser with solid targets. From these first experiments using short pulses (ps) [Clark et al., 2000a; Snavely et al., 2000; Maksimchuk et al., 2000], it therefore appeared that these ion beams presented extremely interesting characteristics. Since, many experiments helped to understand the physics of laser ion acceleration with solid targets, and showed that this source possesses unique characteristics in comparison to beams produced using conventional accelerators:

- Excellent transverse laminarity (more than 100 times better than conventional accelerators [Borghesi et al., 2004; Cowan et al., 2004]).
- Extremely small equivalent source size (μm), characteristic equivalent to the laminarity that allows to obtain an excellent spatial resolution for radiography applications [Borghesi et al., 2002; Mackinnon et al., 2004].
- Very short duration of the source (~ps), allowing to obtain an excellent temporal resolution for pump/probe applications.
- Large energy spectrum (which can be modulated [Hegelich et al., 2006; Schwoerer et al., 2006; Toncian et al., 2006], as can be modulated the divergence [Patel et al., 2004; Toncian et al., 2006]), allowing to obtain in a single radiography, through the different associated times of flight, a movie of the probed phenomena in a single shot.
- Very high current at the source ~kA.
- Intrisic compactness of the acceleration, from 0 to a few tens MeV in a few ten microns (the compactness of the whole "accelerator" is therefore linked to the one of the laser).

It was demonstrated by studying the acceleration physics that the energetic ions observed were accelerated electrostatically from the non-irradiated surface, by fast electrons generated in the interaction zone between the ultra-intense laser and the target, and propagated through the target, as originally predicted [Hatchett et al., 2000; Wilks et al., 2001]. It was also demonstrated that the divergence parameters or the ion energy distribution could be changed using micro-setups triggered by ultra-short lasers [Toncian et al., 2006].

These studies on the understanding of the properties of laser accelerated ion beams have opened original application perspectives in various domains, for fundamental research (study of warm dense matter) and the development of diagnostics for ICF (Inertial Confinement Fusion), as well as for possible medical applications, extremely motivating even if their effective realization is still uncertain (protontherapy for instance):

- Plasma diagnostics for inertial fusion with lasers (analysis of rapidly evolving electric and magnetic fields), analysis of dense objects (static and dynamic like in the case of the compression of a fusion capsule) with resolutions ~ps and ~μm, or for the ignition [Roth et al., 2001; Atzeni et al., 2002; Temporal et al., 2002; Temporal, 2006; Key et al., 2006] of fusion reaction for ICF [Tabak et al., 1994].
- Production of large scale hot and dense plasmas [Patel et al., 2004; Antici et al., 2006], an important aspect for the study of equation of states.
- Injection in conventional accelerators [Krushelnick et al., 2000a; Cowan et al., 2002, Antici et al. 2008b] and evaluation of the potential of laser-accelerated protons for

medical applications (protontherapy [Bulanov et al., 2002a; Bulanov et al., 2002b; Fourkal et al., 2002; Malka et al., 2004] or production of radio-isotopes [Santala et al., 2001; Fritzler et al., 2003; Ledingham et al., 2004; Lefebvre et al. 2006]).

Recently (in 2011), two teams of experimentalists from the Los Alamos National Laboratory beat the maximum laser accelerated ion energy world records one after the other, obtaining 80 MeV and then 120 MeV for protons and Carbon ions with energies higher than a GeV with a 130 TW laser system and an intensity of a few $10^{20}$ W/cm$^2$. The main laser ion acceleration mechanisms are now known and understood, the acceleration by hot electrons (or Target Normal Sheath Acceleration, TNSA), and the acceleration by radiation pressure, but new regimes are explored and an important theoretical and experimental effort is ongoing to further our understanding of laser ion acceleration with various types of targets, to understand its limits in the high laser intensity and in the high laser energy regimes, and to get closer to some applications (Fast Ignition, hadrontherapy...).

Section 2 of this Chapter will present laser ion acceleration with flat dense targets, both through plasma expansion and through the laser radiation pressure with some details on the particular case of circularly polarized laser pulses. The transparency regime is also discussed. Section 3 is devoted to laser ion acceleration using low density targets, either gas jets, exploded foils or foams. Section 4 is devoted to laser ion acceleration using structured targets like cones or reduced mass targets to improve specific characteristics. Section 5 is devoted to the ultra high laser intensity and ultra high laser energy regimes that are starting to be accessible with recent laser systems and that will be further explored in the near future with laser systems under construction. Section 6 is devoted to the presentation of several applications of laser ion acceleration and Section 7 presents the conclusions and perspectives of this Chapter.

## 2. Laser ion acceleration with flat dense targets

Laser ion acceleration using flat solid foils has been the subject of numerous studies in the last ten years, experimental, theoretical and numerical. This Section presents some important results obtained in these studies. First, results on the TNSA regime are presented with the main scaling laws for the maximum proton energy in this regime. Laser ion acceleration through the laser radiation pressure is then introduced with its extension to circularly polarized laser pulses. Both recent theoretical results and first experimental demonstrations are discussed. The special case of the transparency regime is then presented allowing a coupling of the TNSA and of the radiation pressure regime.

When a high intensity laser (>$10^{18}$ W/cm$^2$) with a short pulse duration (30 fs – 10 ps) irradiates a target, this pulse produces, at the effective critical density (~$10^{21}$ cm$^{-3}$), where the laser pulse can not penetrate the plasma, a population of "fast" electrons through different mechanisms (acceleration by the ponderomotive potential [Wilks et al., 1992], resonant absorption [Estabrook et al. 1978], the Brunel effect [Brunel, 1987], mechanisms that are not the primary goal of this Chapter). The dominant mechanism accelerating the electrons depends on the gradient on the front surface and on the laser incidence angle [Lefebvre et al., 1997]; moreover, depending on the interaction conditions, the electrons are accelerated, either in the laser direction, or in the target perpendicular direction [Santala et al., 2000].

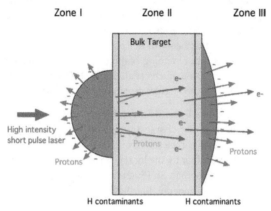

**Figure 1.** Illustration of the various mechanisms leading to high energy ion acceleration by a short and high power laser interacting on a solid target.

In zones I and III (Figure 1), the mechanism is a plasma expansion, with at its front a Debye sheath of hot electrons. In zone III, where there is no initial gradient (in zone I, it is possible only by using very high temporal contrast lasers, which has only been possible recently [Ceccotti et al., 2007]), there is an important difference with zone I: initially, the situation is the one of a virtual cathode where electrons partially extend in vacuum whereas atoms remain structured on the target surface. This is the origin of the high quality of the beam. Following this initial phase, the electrons steadily transfer their energy to the ions and the charge separation field associated to the electron sheath gradually decreases until the acceleration stops, when the electrons are at the same velocity as the ions. This results in the existence of a "cutoff" energy, i.e. a maximum energy of the ion beam, already observed in the seventies. In zone II, protons are accelerated by the strong electrostatic field set up when the laser pulse pushes electrons inside the target.

Except the energy reached by the ions, the main difference between the accelerations in zone I and II on one hand and III on the other hand is that one develops from a pre-existing plasma while the other one originates from an initially cold surface, unperturbed. Therefore, acceleration in zone I or II produces a beam largely divergent ($2\pi$ sr) and turbulent (the structure of the ion front is thermal, not ordered) while acceleration in zone III produces a beam with a limited divergence (determined by the spatial structure of the hot electron sheath), and especially, extremely laminar thanks to the structure of the back surface (if it has a good surface quality) [Fuchs et al., 2005; Fuchs et al., 2007a].

## 2.1. The target normal sheath acceleration

The production of energetic electrons when a high intensity laser interacts with a solid foil has been explained above. These electrons are going to form a hot electron cloud around the target leading to a plasma expansion on both sides of the target [Wilks et al., 2001]. The resulting electrostatic field will then accelerate ions located on both sides of the target to high energies. The amplitude of this field (~TV/m) depends on the hot electron temperature and density. Protons can also be accelerated if they are present in the target or on its surface as contaminants. The basics of the theory of hot plasma expansion [Mora, 2003; Mora, 2005] will be presented and their use in obtaining scaling laws for the maximum proton energy will be explained [Fuchs et al., 2006a; Fuchs et al., 2007a, Robson et al., 2007]. This regime is the so-called "Target Normal Sheath Acceleration" regime (TNSA).

The space charge fields in zones I and III are evidently determined by the local electron density and by the extension of the charge separation [Hatchett et al., 2000; Wilks et al., 2001]. The separation length between electrons and ions is given either by the Debye length ($\lambda_D$~1 $\mu$m for 1 MeV electrons), either by the local gradient length (notably for the part on the front of the target) if it is greater than $\lambda_D$ [Fuchs et al., 2007b]. As in most cases the laser pulse has a pedestal, a plasma is generated at the front surface and therefore the field is in general lower in zone I than in zone III, which confirms the experimental measurements: the energy of ions accelerated towards the laser (in zone I) is smaller than the energy of the ions accelerated in the same direction as the laser propagation direction (in zone II or III) [Ledingham et al., 2004; Yang et al., 2004]. Thanks to the density and temperature of electrons in zone III, and to the value of $\lambda_D$, the field in zone III is initially of the order of a few TV/m.

Acceleration in zone III is the most suited to produce a high quality beam, laminar, that is adapted to imaging applications. Because of the high energy required for electrons to reach zone III, this mechanism was only made possible by the advent of high power laser with short pulse durations. This mechanism also produces, with nowadays lasers, more energetic ions than the acceleration in zone II. The respective energy that can be reached by the accelerated protons in the three zones is illustrated in [J. Fuchs et al., 2007a]. Acceleration in zone II produces less energetic protons for different laser/target conditions as confirmed by PIC simulations performed in a wide range of parameters (for laser intensities going from $10^{17}$ to $5 \times 10^{19}$ W/cm², and for laser pulse duration ranging from 10 fs to 500 fs) [Sentoku et al., 2003; Murakami et al., 2001; Pommier et al., 2003; Pukhov, 2001]. It is also noteworthy that the acceleration in zone II produces a beam with a lower quality [Fuchs et al., 2005] due to the stochastic nature of laser-plasma interaction (in the preplasma) which does not produce a charge separation field structure at the critical density interface as ordered as the one existing at the back (non perturbed) surface of the target (zone III). The collimation and low emittance of rear side accelerated ions are considerable assets for numerous applications.

Nevertheless, we can note that the use of high temporal contrast lasers will erase the differences between the acceleration in zones I and III: in these conditions, there will be no preplasma, and therefore the spatial quality of the ion beams produced in zones I and III will be similar [Ceccotti et al., 2007]. If the target is, in addition, very thin, the electron populations will be similar on both sides of the target and the ion energy will therefore also be very close. On the contrary, for thick targets, the electrons will have a lower density when arriving at the back surface of the target (zone III), and the resulting accelerating field as well as the maximum ion energy will thus be lower in this zone than in zone I.

An acceleration model has been developed to obtain a predictive model of maximum ion energy useful for experimentalists [Fuchs et al., 2006; d'Humieres et al, 2006; Fuchs et al., 2007a]. This model uses theoretical results on the isothermal expansion of a hot plasma [Mora, 2003] coupled to a model of characteristic ion acceleration time. By studying the dependence of this characteristic time on various parameters, it was possible to extend the validity of this model to very short pulse durations.

This study has also allowed to link the characteristics of the TNSA accelerated ion beams to the characteristics of the energetic electron population in order to develop a new diagnostic of this electron population for laser-plasma interaction experiments [Antici et al., 2008a].

To obtain an analytical estimate of the maximum energy that can be gained by protons accelerated forward from the back surface of the target (zone III) [Fuchs et al., 2006; d'Humieres et al, 2006; Fuchs et al., 2007a] it is possible to use the approach chosen by Mora [2003], which was based on an approach initiated in the seventies to treat the expansion of a plasma in vacuum, and adapted to the case of a hot electron burst initiating the expansion. In the case of a self-similar isothermal expansion [Gurevitch et al., 1966], a simple evaluation of the accelerated ions maximum energy is:

$$E_{max/back} = 2ZT_{hot} * [\ln(t_p + (t_p^2 + 1)^{1/2})]^2$$

Where $T_{hot}$ is the electron temperature in the sheath in zone III and $t_p = \omega_{pi} * t_{acc} / (2 * \exp[1])^{1/2}$

with $t_{acc}$ the acceleration time and $\omega_{pi} = \sqrt{\dfrac{n_{e0} Z e^2}{M_i \varepsilon_0}}$ the ion plasma frequency ($n_{e0}$ is the initial

electron sheath density at the back of the target and $M_i$ the ion mass). As already mentioned, in most experiments, a plasma is located on the front surface during the target irradiation by the main pulse. In this case, as already showed in experiments, $T_{hot}$ can be well approximated using the ponderomotive potential $m_e c^2(\gamma - 1)$ [Malka et al., 1996; Popescu et al., 2005]. In reality the expansion is far from being isothermal, and is actually adiabatic [P. Mora, 2005]. Starting from a hot electron cloud, there is a progressive energy transfer through the electrostatic fields to the ions. Nevertheless, such an adiabatic model cannot be solved analytically, only numerically.

The ambition of this model was to see if the isothermal model could be used, having the advantage of being simple and easy to use analytically, by compensating with a finite (determined) acceleration time the fact that the hot electron population is maintained at a constant temperature [Fuchs et al., 2006]. To confront this approach with experimental results and see how interesting it is, it is important to determine the missing ingredients, namely $n_{e0}$ and $t_{acc}$.

To determine $n_{e0}$, the quantity of hot electrons accelerated inside the target by the laser has been evaluated in a simplified way, with an energy balance. This quantity can be evaluated as $N_e = f\ E/T_{hot}$, where $f$ is the laser energy fraction absorbed in the plasma and converted in hot electrons. This fraction $f$ was measured by using a buried layer inside the targets and by measuring the X-ray radiation induced in these layers by the hot electrons. This way, it is possible to determine the total energy in the hot electrons and therefore $f$. It depends of the incident laser intensity: $f = 1.2 \times 10^{-15} I^{0.74}$ (where I is in W/cm² units) with a maximum $f$ of 0.5. We can now assume that these electrons are, longitudinally, confined and transversally diluted in the target [Kaluza et al., 2004] until they are spread on the electron sheath area $A_{sheath}$ at the back of the target. This would give $n_{e0} = N_e/(d \times A_{sheath})$ where $A_{sheath} = \pi(r_0 + d \times \tan\theta)^2$, d is the initial target thickness and $r_0$ is the radius of the zone from which the electrons are accelerated at the critical interface, radius that we assume to be equal to the focal spot radius ($r_0 = FWHM$). $A_{sheath}$ also depends on the hot electron divergence half-angle in the target $\theta$. Several experimental and theoretical studies show that $\theta$ is between 15° and 40° [Fuchs et al., 2003; Stephens et al., 2004; Adam et al., 2006], increasing with laser energy [Lancaster et al., 2007; d'Humières et al., 2006].

In the model, $\theta = 25°$ has been chosen in agreement with what the LULI team had deduced from astigmatic proton emission measures [J. Fuchs et al., 2003]. This model is supported by the fact that:

- The values of the size of the electron sheath at the back of the target obtained with this model are in good agreement with direct measures that can be obtained of the size of the proton source.
- The values of electron density at the back of the target as well as the ones of the zone on which these electrons spread are also in good agreement with direct measures of these parameters performed using an interferometry diagnostic with a probe beam reflected on the back of the target that was adapted to this problem [Antici et al., 2008a].

To determine the accelerating time $t_{acc}$, different approaches were tested by comparing them not only to data obtained during experiments for laser intensities ranging from $2 \times 10^{18}$ to $6 \times 10^{19}$ W/cm², and for laser pulse durations between 150 fs to 10 ps [Fuchs et al., 2006], but also to data obtained with shorter pulse durations taken from various articles. A priori, one could think that $t_{acc}$, which represents physically the time during which in the model the electron temperature is maintained to $T_{hot}$ must be proportional to the laser pulse duration. Nevertheless, for very short pulse durations, it is important to take into account a minimum time for the energy transfer between electrons and ions. Therefore, for very short pulse durations, the acceleration time is not proportional to the laser pulse duration anymore but needs to tend towards a constant value. It is also necessary to include in the model the fact that this acceleration time needs to depend on the laser intensity. Indeed, for low laser intensities, this acceleration time needs to be increased because the expansion is slower. In the end, the function, which is most adapted to the different experimental measures, is [J. Fuchs et al., 2007a]: $t_{acc} = \alpha^*(\tau_L + t_{min})$ where $\tau_L$ is the laser pulse duration, $t_{min} = 60$ fs and where $\alpha$ goes linearly from 3 for an intensity of $2 \times 10^{18}$ W/cm$^2$ to 1.3 at $10^{19}$ W/cm$^2$ and is constant at 1.3 for higher intensities. The figures in [Fuchs et al., 2006; d'Humières et al., 2006], as well as Figure 2 show a comparison between the calculations of the maximum proton energy $E_{max/back}$ and experimental results obtained in a wide range of laser and target parameters (in particular for short pulse durations, and/or low intensities that were not explored experimentally at LULI). This model was also validated using 2D PIC simulations [d'Humières et al., 2006].

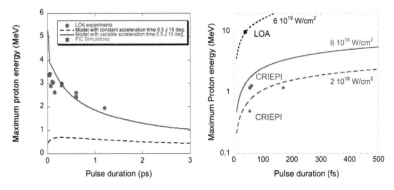

**Figure 2.** (Left) Data (points), simulations (squares) and model (lines) giving the maximum proton energy as a function of the pulse duration for very short pulse durations. In the model, a focal spot of 6 μm FWHM is used as well as a target thickness of 5 μm. The different lines correspond to different intensities for which the experiments were performed. (Right) Same graph but for various laser systems (points) and the model (lines). Here a thickness of 5 μm was used in the model as most of the reported experiments used a similar thickness.

It is nevertheless important to keep in mind that the acceleration time $t_{acc}$ is not the « true » acceleration time, it is the typical time during which the electron population remains at a « high » temperature. In reality, the hot electron temperature decreases progressively during the energy transfer between electrons and ions.

To conclude this Section, the analytical work on the maximum proton energy produced with the accelerating mechanism in zone III has allowed to obtain an evaluation tool that is able to correctly reproduce the existing data obtained at LULI and other laser systems, and to extrapolate it to other laser parameters (it is important to keep in mind that this model can only be entirely reliable in the parameter range in which it has been tested).

Effects of a small scale gradient at the back of the target on laser ion acceleration in this regime has also been investigated [Fuchs et al., 2007b], and the physics of long pulse interaction and its effects on the creation of the accelerating electron sheath is now being intensely studied.

## 2.2. Laser ion acceleration through the laser radiation pressure

Laser ion acceleration through the laser radiation pressure has first been studied theoretically and using Particle-In-Cell simulations by Wilks et al. [1992] and Denavit et al. [1992]. It was first referred as the hole boring regime and can lead to the development of strong electrostatic shocks inside the target [Silva et al., 2004]. Electrons are pushed by the laser ponderomotive force at the front of the target and lead to a strong charge separation setting up a high amplitude electrostatic field. This field will accelerate front surface ions to high energies during the duration of the laser pulse. This regime is also referred to as Radiation Pressure Acceleration (RPA) or the laser piston regime.

It is possible to estimate analytically the maximum proton energy of forward accelerated protons from the front surface. Indeed, the maximum velocity of the accelerated protons in zone II is given by two times the receding plasma surface velocity [Pukhov, 2001; Denavit, 1992]. In the case of a total laser reflection at the critical interface (i.e. when the laser piston efficiency is maximum), this receding velocity is given by [Wilks et al., 1992; Wilks, 1993] :

$$\frac{u_s}{c} = \sqrt{\frac{1}{2}\frac{n_c}{n_e}\frac{Zm_e}{M_i}a_0^2}$$

Where $n_c$ is the effective critical density, where the laser wave reflection takes place, i.e. $n_e=\gamma n_c$ [Fuchs et al., 1999] at normal incidence with $\gamma$ being the Lorentz factor given by $\gamma=(1+a_0^2)^{1/2}$, with $a_0^2=(p_{osc}/mc)^2=I\lambda^2/(1.37\times10^{18})$, I being the laser intensity (in units of $W/cm^2$) and $\lambda$ is the laser wavelength in microns.

Therefore:

$$E_{max/front} = 2*M_i*u_s^2=Zm_ec^2a_0^2/\gamma$$

One can note that this evaluation is consistent with the values measured specifically for protons produced in zone II [Kaluza et al., 2004] that can indeed be measured using two different techniques:

- With neutron spectroscopy: by using a $CD_2$ target and by measuring the emitted neutrons [Youssef et al., 2006; Habara et al., 2004].
- With nuclear activation [Nemoto et al., 2001]: a thin layer containing Deuterium is placed at the front of Al targets. Experimentalists were able to quantitatively determine the $D^+$ spectrum emitted in zone II for which the maximum cutoff energy compared favourably with the formula above [Fuchs et al., 2005; Fuchs et al., 2007a].

The analytical work on the maximum proton energy produced with the accelerating mechanism in zone II has allowed showing that the acceleration mechanism in zone III produces, in almost every case, the maximum proton energies [Fuchs et al., 2007a].

This conclusion can be obtained using four different means:

- First, experimentally, it was possible to demonstrate that acceleration in zone III produces the forward accelerated protons with the maximum energy.

- Then, the same conclusion related to the fact that acceleration in zone III is dominant can be reached by observing the agreement between the experimental measures and the predictions of the acceleration model in zone III (see above).
- The models described in this Section and in Section 2.1 are also in good agreement with 2D PIC simulations over a wide range of target and laser parameters.
- Finally, when comparing systematically the maximum proton energies obtained by the two mechanisms, one also notes the dominance of acceleration in zone III to generate the most energetic protons. This comparison was done for various pulse durations, target thicknesses and laser intensities [Fuchs et al., 2007a]. The dominance of zone III is systematic, except for ultra-short pulse durations (< 10 fs) for which the laser energy is too low and the number of electrons accelerated towards the back surface is therefore low, thus reducing the amplitude of the accelerating field at the back of the target.

With this study, the controversy on the origin of the most energetic protons was concluded.

More recently, this regime has been revisited in the case of a circularly polarized laser pulse. In this case, a strong electrostatic field is also created at the front of the target as electrons are pushed forward but they are not heated. They therefore create a thin electron layer travelling inside the target. The hot electron temperature and density are therefore decreased, also decreasing the efficiency of the TNSA mechanism. The thin accelerated electron layer will drive a strong charge separation that can lead to the acceleration of quasi-mono-energetic ion beams. Recent numerical and theoretical studies [Klimo et al., 2008; Robinson et al., 2008; Macchi et al., 2009; Grech et al., 2011; Schlegel et al., 2009; Tamburini et al., 2010], as well as recent experimental studies [Henig et al., 2009] have explored this regime. It can also be explored using higher wavelength with $CO_2$ lasers interacting on gas jets [Palmer et al., 2011; Haberberger et al., 2011].

## 2.3. Laser ion acceleration in the transparency regime

A new regime of laser ion acceleration which relies on the interaction of high contrast laser pulses with ultra thin foils has been validated using simulations [Dong et al. 2003; d'Humières et al., 2005; Yin et al., 2006; Albright et al., 2007] and experiments [P. Antici et al., 2007; D. Neely et al., 2006]. In this regime, the laser is able to go through the target during the duration of its interaction. Both the TNSA mechanism and the radiation pressure acceleration can play an important role in this case and laser energy absorption is increased as volume absorption occurs.

It has been shown that proton energy could be increased by reducing the solid target thickness [Mackinnon et al, 2002; Kaluza et al., 2004]. For extremely thin targets where relativistic transparency for the laser pulse occurs, theoretical studies [Dong et al., 2003; d'Humières et al., 2005; Esirkepov et al., 2006; Yin et al., 2006] have suggested that proton acceleration could be even more efficient than using the standard regime of acceleration from solids [Allen et al., 2004; Fuchs et al., 2005]. Antici et al. [2007] report on some of the first experiments performed to explore the regime of ion acceleration using ultra-thin targets interacting with an ultra-high temporal contrast 320 fs duration laser pulse. Proton beams accelerated to a maximum energy of ~7.3 MeV from targets as thin as 30 nm thick and for a $10^{18}$ W.cm$^{-2}$ laser intensity were observed. Neely et al. have also reported [2006] results using similar targets in ultra-high contrast conditions. Here, using a much lower on-target laser intensity and a longer pulse ($10^{18}$ W.cm$^{-2}$ and 320 fs) than Neely et al. ($10^{19}$ W.cm$^{-2}$ and 33 fs), ~2 times higher proton energies were obtained. It was also shown that this acceleration regime produces nearly a tenfold increase of proton maximum energy as compared to the standard regime using thicker targets at the same laser intensity. PIC simulations in close agreement with the measurements suggest that enhanced target electron heating as compared to thicker targets results in the observed proton energy increase. In the simulations much higher proton energy increase is predicted for similar target thickness and somewhat higher laser intensities (3-6 $10^{19}$ W.cm$^{-2}$).

As in the rear-surface acceleration process with thick targets, ion acceleration in the relativistic transparency regime involving ultrathin targets is observed in simulations to be lead by the fast electrons accelerated by the laser from the front side into the target, except that a higher number of electrons are heated to higher temperatures as the result of the interaction of the laser pulse with the whole volume of the ultra-thin target. Emerging from the target, these electrons form a dense electron plasma sheath at the vacuum interface that accelerates surface ions [Hatchett et al., 2000], mostly hydrogen-rich contaminants [Gitomer et al., 1986].

In the standard regime of thick targets, the proton energy increase for decreasing target thickness is due to a reduced spreading of the fast electrons within the target as discussed in Section 2.1. For the relativistic transparency regime, an optimum target thickness is predicted to be reached when the laser absorption in the target equals the laser transmission [d'Humières et al., 2005], in practice it is in the range of tens to hundreds of nm. At the peak intensity levels (> $10^{18}$ W/cm$^2$) required to accelerate ions to large energies, the pedestal pre-leading the main pulse, or ASE (laser amplified spontaneous emission), is generally high enough to ablate such thickness of target material. More importantly, it also induces rear-surface target heating prior to the arrival of the main pulse for thin targets (still thicker than the ablated thickness), reducing the ion energy. This has limited the minimum target thickness to few microns in typical high-energy ion acceleration experiments. This is why laser pulses need to be contrast enhanced in order to explore this relativistic transparency regime. The ultrahigh contrast can be achieved using several methods like self-induced plasma shuttering from solids [Doumy et al., 2004], or frequency doubling of the laser wavelength. Both induce similar reduction in laser energy but the latter has the further disadvantage of decreasing the $I\lambda^2$ factor.

A maximum proton energy of ~7.3 MeV was recorded during these experiments using two plasma mirrors to be able to use thinner targets. The targets were 30 nm thick SiN planar membranes uncoated or coated with a variable thickness of Al. Without plasma mirrors, the laser interaction with ultra-thin targets yielded no result (probably because the target material was evaporated before the main pulse) and the laser interaction with 5 μm thick targets also yielded a maximum energy below the detection threshold, i.e. <0.9 MeV. At similar laser intensities, the proton energy increase going from thick target with a standard contrast to ultra-thin targets with ultra-high contrast was thus higher than a factor 8. Equivalently, to reach similar proton energies in the standard regime of thicker target having cold rear-surfaces, much higher laser intensities are needed, e.g. 2 $10^{19}$ W.cm$^{-2}$ for 25 μm thick targets [Fuchs et al., 2006]. Laser to proton energy conversion, calculated by considering protons from 1.8 MeV (resp. 4 MeV) to the maximum energy is 4 % (resp. 1.1 %) for 30 nm thick target, 3.7 % (resp. 0.9 %) for 80 nm thick target, and 2 % (resp. 0.4 %) for 100 nm thick target. Note that when the same maximum proton energy is reached in the regime of thicker targets at much higher laser intensity, the laser to proton energy conversion is much lower, i.e. ~0.1 % for protons from 4 MeV to the maximum energy. The angular distribution of the proton beams is also observed to be irregular, on the contrary to proton beams accelerated at similar maximum energies from thick metal targets which display a very smooth profile. This could be due to the fact that the focal spot in presence of two plasma mirrors is not optimized.

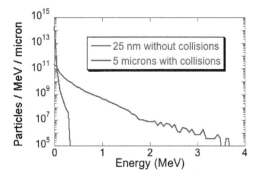

**Figure 3.** Electron spectrum obtained from the 2D PIC simulations for a 25 nm thick target and a 5 microns target, a 350 fs pulse duration, and a 1.38×10$^{18}$ W/cm$^2$ laser intensity.

Simulations performed using the 2D PIC code PICLS [Sentoku and Kemp, 2008] of the interaction of a prepulse-free laser pulse with a thin dense target allow to get an insight on the interaction conditions. The incident laser pulse has a 1 μm wavelength, a 350 fs pulse duration, an intensity of 1.38×10$^{18}$ W/cm$^2$, and a transverse spot size on target of 6 μm FWHM. The temporal and spatial (transverse) profiles are Gaussian and the laser electric field is in the simulation plane. The laser interacts at normal incidence with the 25 nm thick, 100 $n_c$ density target, initially fully ionized, that consists of a layer of Al surrounded by two layers of hydrogen.

The maximum proton energy is 7.33 MeV, as observed 2.8 ps after the beginning of the simulation, in close agreement with the observation. Performing a simulation with the same parameters but adding the treatment of collisions yields a very close maximum proton energy of 7.28 MeV, as can be expected since in such conditions where the target is heated efficiently, the jxB absorption significantly dominates collisional absorption. In both simulations, we observe negligible transmission. The consistency between experiment and simulations support the fact that in the experiment the main laser pulse could interact with a solid density target.

To understand the specific mechanism of proton energy increase obtained using ultrathin targets, an additional simulation using a thicker, i.e. 5 μm thick, target still at 100 $n_c$, with collisions was performed. These simulations show a higher heating for ultrathin targets compared to the thicker target. We have to note that the laser absorption is comparable for the two targets (~23 %). It is the partition of the absorbed laser energy into the cold and hot electron populations that differs significantly between the two target thicknesses, as shown in Figure 3. In the thicker target, a high number of electrons remain cold and the maximum electron energy is low. In the ultrathin target, the hot electrons spread around the whole target and the maximum electron momentum is high. This is due to the target heating induced by the hot electrons, which are not damped by the cold background as in the thicker target, leading to a density lowering of the target, and an enhanced absorption in hot electrons for subsequent parts of the laser pulse.

Interaction with ultrathin targets leads to a very efficient electron heating, and high proton energies. Going towards longer pulse durations also favors absorption in hot electrons in ultra-thin targets since the trailing part of the pulse can take advantage of the target heating induced by the leading part and of the increased absorption. This explains the increased proton energy observed in these experiments using 320 fs duration pulses as compared to experiments performed with 30 fs duration pulses [Neely et al., 2006], although the intensity used here is ten times lower.

The experiments and simulations presented here show the potential of laser-acceleration of protons from ultra thin targets using ultra-high contrast pulses. In the present experiment, exploring of this regime has been limited to the threshold of the relativistic regime. Both the low efficiency of the consecutive plasma mirrors and the low final focal spot quality are strong limitations. In future studies, this could be improved by adjusting the fluence on the mirrors. It has been indeed demonstrated that 50 % efficiency with good focusing quality could be obtained from two successive mirrors [Wittmann et al., 2006]. This would allow achieving intensities above $10^{19}$ W/cm² , thus working with an optimum configuration for the same target thicknesses. In this case, as shown in [Antici et al., 2007], 2D simulations show that very high proton energies (> 100 MeV) should be obtained. Around 120 MeV protons and GeV Carbon ions were recently measured in this regime at the Los Alamos National Laboratory [Jung, 2012]., breaking the existing world records, and plasma mirrors are now routinely used in laser ion acceleration experiments with ultrathin targets.

## 3. Laser ion acceleration using low density targets

In this Section experimental and theoretical studies on laser ion acceleration with low density targets are presented. This regime offers a promising alternative to the accelerating regimes described in Section 2. The basics of this regime are described and some recent experimental and theoretical results are also discussed.

As discussed in sections 1 and 2, intense research is being conducted on sources of laser-accelerated ions and their applications, e.g. radiography and production of Warm Dense Matter (see review articles [M. Borghesi et al., 2006; J. Fuchs et al., 2009]). This is motivated by the exceptional properties that have been demonstrated for proton beams accelerated from planar targets (see Section 2), such as high brightness, high spectral cut-off, high directionality and laminarity, and short duration (~ps at the source). In most experiments, solid density targets, typically gold or aluminum, are used and the most energetic ions are accelerated from the rear side by a strong electrostatic field created by fast electrons (through the so-called TNSA mechanism). This section presents a different setup aiming at studying the possibility of enhancing the efficiency and ease of laser acceleration of protons and ions compared to what has been achieved up to now using standard TNSA.

Indeed, it is not only important for applications of such ion beams that their parameters (maximum energy and total number) are enhanced, but also that they are easily usable. Regarding the second point, there are currently some issues/limitations linked with using solid targets: 1) targets need to be aligned precisely for each shot, 2) laser temporal contrast needs to be controlled, 3) debris are produced, 4) repetition rate is limited. These limitations would all be greatly alleviated if one could use gas jets as the laser interaction medium. Following preliminary theoretical [Esirkepov et al., 1999; Yamagiwa et al., 1999; Sentoku et al., 2000] and experimental [Krushelnick et al., 1999; Sarkisov et al., 1999] studies on ion laser acceleration using underdense targets, ion acceleration from low density targets was recently the subject of growing attention [Willingale et al., 2006; Yogo et al., 2008; Antici et al., 2009; Willingale et al., 2009]. However, the efficiency of the process was not shown to be high. For example in [Willingale et al., 2006], the maximum energy of the ions produced this way was only half of the one that could be obtained using solid targets.

There are several interpretations about the (best) production of the proton beam in gaseous conditions. One hypothesis is that the protons are produced by electrostatic acceleration at the target–vacuum boundary like in TNSA, although this was subject of debate [Willingale et al., 2007]. For example, acceleration of energetic ions observed by Willingale et al. had been explained in terms of strong inductive electric fields due to magnetic fields variations on a steep density gradient [Bulanov et al., 2007]. Through simulations, it has been found that shock acceleration could be used in such low-density medium to accelerate ions not only very efficiently, but also with a comparable number as TNSA [d'Humières et al., 2010a]. The basic idea behind it is that lower density targets can improve the absorption of the laser energy, hence both the laser-to-ion conversion and the ion energies can be enhanced. However, to reach such optimum working point, precise interaction conditions need to be met. With present-day laser parameters, it requires rather thin gas jets (of the order of 100 microns), which are not readily available. So, in parallel of working to produce such gas jet, it was proposed to work on demonstrating such optimum of ion acceleration exploiting shock acceleration in lower-than-solid density targets [d'Humières et al., 2010b]. For this, it is possible to use, as a substitute for optimized gas jet, targets that have been exploded by a secondary laser prior to the arrival of the main beam. The goal is not only to demonstrate the existence of optimized ion acceleration beyond TNSA, but also to shed light about these processes.

Test experiments have recently been performed at LULI (France) demonstrating the effectiveness of ion acceleration in the regime which had been explored through simulation. However the laser intensity and energy during that experiment were somewhat limited (~J) [Gauthier et al., 2012]. As PIC simulations (see below) show that this acceleration process is more efficient with higher intensities and longer pulse durations, doing experiments on that topic using higher energy lasers would be very beneficial as it would allow to effectively test the increased efficiency of shock acceleration in the domain where it is seen by simulations to be best: with long pulses and high laser energy. This, coupled to progresses on gas jets, would strongly enhance the practicality and possibilities for ion probing or warm dense matter generation on future high-power and high rep-rate laser facilities.

As found out by 1D and 2D Particle-In-Cell simulations performed with the PICLS code with respect to previous experiments using exploded foils having low-density [Antici et al., 2009], the ion acceleration processes in these plasmas depend strongly on the characteristics of the density gradient. For sharp and small density gradients, the most energetic protons are accelerated at the back surface, similarly as in TNSA. The acceleration mechanism is then similar to the one observed with solid foils. For intermediate density gradients, the most energetic protons are accelerated by a collisionless shock mechanism in the decreasing density ramp at the target exit. A two-step acceleration process takes place: first, ions are accelerated in volume by electric fields generated by hot electrons, second, the ion energy is boosted in a strong electrostatic shock, as shown in [E. d'Humières et al., 2010a]. If the density profile becomes too long, the shock cannot be triggered and ions are accelerated by normal expansion in the density gradient which is a low efficiency mechanism [Grismayer and Mora, 2006].

Collisionless shocks have already been studied in decreasing density gradients for spherical plasmas [Peano et al., 2007] and for plasmas located at the back of a solid foil irradiated by a laser [Tikhonchuk et al., 2005]. In [d'Humières et al., 2010b], and [E. d'Humières et al., 2011a; E. d'Humières et al., 2011b], the shock regime and the two step process were studied in detail. The first step, the launch of a fast ion wave, requires a hot electron population and a descending density profile and occurs in a zone of high amplitude magnetic fields. This first step has already been described theoretically by Grismayer and Mora [2006] but since the authors used a Lagragian code, they could not go beyond this first step. The second step, the development of a strong electrostatic shock which boosts the energy of the ions, happens when the ion bunch resulting from the first step enters a low density plasma region where the magnetic field has strongly decreased. The first 3D simulation of this regime were recently performed [d'Humières et al., 2011a] and confirmed the 2D simulation results. Using 1D PIC simulations, it was also confirmed that this regime does not require a high magnetic field to launch a strong collisionless shock. A Vlasov-Poisson code is presently used to study in more details the processes described and to obtain scaling laws of the maximum proton energy and accelerated proton numbers with laser and target parameters [d'Humières et al., 2011b]. The number of accelerated protons is high, similar to what is observed with TNSA, and could be higher as it is possible to accelerate ions in a thicker region of plasma even if the density is low in this region, whereas TNSA is limited to a small region inside the back of a target where the strong electrostatic field can be felt. As exemplified by Figure 4, 2D PIC simulations show that to reach this high efficiency regime, a high laser energy is needed for low density plasmas with a high thickness by density product. The fact that it is difficult nowadays to produce these types of plasmas except from exploded foils naturally gives an advantage to large-scale facilities with several tens of Joules in the main pulse and several tens of Joules in the exploding pulse.

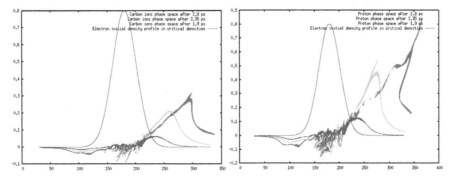

**Figure 4.** (right) Proton phase space at three different times (1.9 ps after the beginning of the simulation for the blue dots, 2.35 ps for the green dots and 2.8 ps for the red dots) and initial electron density profile (purple line, units of critical density). The laser is injected from the left. The target is composed of Carbon ions, protons and electrons and its FWHM is 50 microns. Laser intensity is $5 \times 10^{20}$ W/cm$^2$, pulse duration is 700 fs FWHM and the focal spot width is 6 microns. Maximum proton energy at t=2.8 ps is 281 MeV. It reaches 296 MeV after 3 ps. The development of the ion wave is clearly visible at t=1.9 ps and the shock is launched just before t=2.35 ps and has fully developed at t=2.8 ps. (left) Carbon ions phase space at three different times (1.9 ps after the beginning of the simulation for the blue dots, 2.35 ps for the green dots and 2.8 ps for the red dots) and initial electron density profile (purple line, units of critical density). The Carbon ion wave is clearly visible at t= 1.9 ps and 2.35 ps and the shock on Carbon ions starts to develop just before t=2.8 ps. This is later than for the shock on protons. This is normal as the Carbon ion wave takes more time to develop than the proton wave. Maximum Carbon ion energy is already 496 MeV at t= 2.8 ps. It reaches 889 MeV after 4 ps. In both cases, the x-axis is in microns and the y-axis for the phase spaces represents beta*gamma where beta is the ion velocity divided by the velocity of light and gamma is the proton Lorentz factor.

As illustrated in Figure 4, preliminary simulations using the TITAN (LLNL) laser parameters were performed. These show the extremely interesting potential of this regime since in the simulations 296 MeV protons and ~900 MeV C ions are accelerated. Other simulations using TITAN parameters but exploring different target thickness of plasmas (still peaking at 0.8 $n_c$) have been performed. For all cases, the production of energetic particles is linked to strong electrostatic shocks. More simulations and experiments are needed to fully grasp the potential of this regime and to optimize it.

## 4. Optimization of laser ion acceleration with various types of targets

In this Section, several innovative studies on the optimization of the characteristics of laser accelerated ion beams are presented. New targets were recently tested to change the interaction conditions, the electron sheath characteristics or the ion distribution. These studies have allowed to find new ways to increase the maximum ion energy, as well as new ways to decrease the beam divergence and energy spread. Several of these targets or setups are described here. Other targets have recently been investigated to improve the coupling of the laser energy with the target electrons but are not described in this Section. Klimo et al. [2011] have for instance tested nano-structured targets to enhance the coupling between the laser and energetic electrons.

### 4.1. Laser ion acceleration using micro-cones

Studies of the interaction of energetic laser pulses with micro-cones using PIC simulations have been performed in collaboration with experimentalists from Los Alamos National Laboratory and the University of Nevada, Reno. The importance of the geometry of the cone on the increase of the laser energy absorption and of the hot electron temperature was highlighted. It was also shown that these targets are very sensitive to the laser alignment and contrast. Coupling these cones with flat micro-disks, it is possible to strongly increase the maximum proton energy [Flippo et al., 2008] while also increasing the obtained ion beam divergence [Renard Le Galloudec et al., 2010].

The basics of laser ion acceleration with micro-cones are detailed in this Section as well as recent theoretical, experimental and numerical results. The world record in maximum laser accelerated protons energy was broken in 2009 using these types of targets [Gaillard et al., 2011]. A proton beam with energies as high as 69 MeV was obtained using 80 J of laser energy whereas the previous record had remained at 58 MeV [Snavely et al., 2000] for almost 10 years and needed 500 J.

Cone targets are of interest for their potential to increase the hot electron temperature and population density [Sentoku et al., 2003a; Kodama et al., 2004], which are the main contributors to the efficacy of the TNSA mechanism. Sentoku *et al.* [2003a] showed that sharp tip cones can effectively increase the number of electrons available for laser heating while guiding the laser light along the cone wall surface toward the cone apex. This action tremendously increases the interaction area of the laser, producing more electrons and concentrating the laser field at the cone neck near the flat-top surface. UNR physicists in combination with the nano-fabrication group, NanoLabz, were able to design a relatively inexpensive process for mass producing a new type of cone target (Figure 5.a). A process with obvious advantages over the hand-assembled ICF targets to date. Early results have shown these cones can produce excellent ion beams when properly aligned.

Pioneer experiments performed on the Titan laser (LANL) showed that the flat-top cone produces more protons at higher energies than the flat foil, in this case 13 times more above 10 MeV [Flippo et al., 2008]. The total energy present in the beam from the Au flat-foil is measured to be 0.4% of the incident laser energy. When this is done for the flat-top cone target it is measured to be 1.9% of the incident laser energy. This represents a nearly 5 fold increase in the conversion efficiency over the Au flat foil targets and a 3.4 fold increase in the total amount of protons, with nearly 13 times the number above 10 MeV. The proton beam observed from the cone target contains two orders of magnitude more protons than previous experiments at similar intensities [Maksimchuk et al., 2004] and is more than two orders of magnitude more efficient (laser energy conversion to protons) than previously published work for a similar laser intensity [Fuchs et al., 2005] and energy [Fuchs et al., 2005; Borghesi et al., 2006]. The beam also has a 3 to 5 times higher maximum proton energy than previously reported for a similar intensity from flat foils [Fuchs et al., 2006; Robson et al., 2007], at least 1.5 times that of the Trident flat foil, and potentially more than 2 times that of the flat foil as simulations indicate.

The cone targets have been modelled with the particle in cell code PICLS. One simulation was performed exactly matching the experiment at a laser wavelength of 1 μm with an intensity of $10^{19}$ W/cm$^2$, a pulse duration of 600 fs and a focal spot of 12 μm. The cone inner neck diameter was 10 μm, and the flat-top thickness was 13 μm with a diameter of 90 μm. Maximum proton energy reached almost 40 MeV (Figure 5.b). Several other simulations where performed as part of a parameter study at a laser wavelength of 1 μm and an intensity of $10^{19}$ W/cm$^2$, but with a pulse duration of 350 fs and a FWHM spot of 6 μm to keep the calculation time to a minimum. The various simulations are summarized in [Flippo et al., 2008]. The cone parameters have been varied to highlight the main advantages and drawbacks of the flat-top micro-cones and to understand the driving factors behind the observed enhancement. They reveal that the cone leads to significantly improved laser energy absorption and conversion to hot electrons when well aligned, resulting in higher hot electron temperatures and densities. These increases due to the flat-top cones, which are a function of the cone neck-diameter and opening angle [Nakatsutsumi et al., 2007], provide a potential advantage for laser-ion acceleration as these are the main parameters governing the TNSA accelerated ions [Wilks et al., 2001].

At these laser parameters (1 μm laser wavelength, 350 fs pulse, 6 μm spot size, and $10^{19}$ W/cm$^2$ intensity), the flat-foils have been observed experimentally to produce ~3-5 MeV [Fuchs et al., 2006a] for a 20 μm thick target. The simulated flat-foil, which is 10 μm thick, results in a proton maximum energy of 8.6 MeV. This higher energy is consistent with the fact that the thinner simulated target follows the experimental trend of increasing proton energy for a thinner target at the same laser parameters [Mackinnon et al., 2002; Fuchs et al., 2006a]. The smooth cone with a 10 μm inner diameter neck and 10 μm thick walls (for a total neck outer diameter of 30 μm) has the highest performance, yielding a maximum proton energy of ~26 MeV, with a 90 μm top and a top-to-neck ratio of 3. Comparatively, a cone with a 20 μm inner diameter (total of 40 μm) with a top to neck ratio of 2.25 yields 10 MeV, a 62% decrease. Here, when the neck size is increased by a factor of two the electron temperature, the maximum electron energy, and the proton maximum energy are all decreased by about the same factor. Combining information from [Flippo et al., 2008], one can deduce that the neck diameter has a stronger influence on the maximum proton energy and electron temperature than the flat-top diameter. However, it is evident that both dimensions play a role; and thus, their ratio is still a good figure of merit.

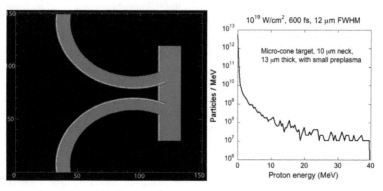

**Figure 5.** (a) Illustration of the shape of the flat-top cone geometry. Axis are in microns (from 0 to 150 microns). (b) Simulated proton energy spectrum.

One can conclude that for the TNSA mechanism to work efficiently, a smooth surface on which a dense, hot sheath can form is essential. A sharp-tip cone would not be suited for this application, and results have not shown promise for such cones to be used as proton beam generators. One also needs to transport the many hot electrons efficiently to the flat-top surface from wherever the hot electrons are generated. This last point can be described as the hot electron population transport characteristic of the cones. The transverse pointing accuracy can have a major effect on this transport. The farther the laser is focused away from the cone axis, the longer and more complex the path is to the flat-top for the hot electrons. On the way, these electrons can be scattered and lost to other areas of the cone or slowed down by the transfer of energy to the ions present on the rear surface of the cone side-walls. The overall hot electron population and temperature is shown to be only slightly affected by the increase in transverse offset distance; whereas the maximum proton energy is largely perturbed, decreasing by a factor of eight. This is caused by the electrons inability to be efficiently transported to the flat-top, and instead spread out over the relatively large surface area of the cone's side-walls [Flippo et al., 2008].

Another issue affecting the hot electron population, temperature, and transport is the preplasma filling in the inner cone neck. As long as the laser is able to propagate through the preplasma without significant filamentation, a higher density preplasma can enhance the coupling to the hot electrons, which can effectively lead to an increase in the maximum ion energy. However, if the laser is strongly affected by a denser preplasma and is instead reflected, dispersed, or entirely absorbed far from the flat-top, then any hot electrons generated must travel a longer distance to arrive at the flat-top. As in the case of the transverse pointing accuracy, the hot electrons will be scattered or decreased in energy traversing the longer path, or completely lost to forming the cone side-wall sheath, never making it to the flat-top. As the density of the plasma inside the cone neck increases the distance between the interaction zone and the flat-top increases. When the level of preplasma density inside the cone neck is increased from $0.001n_c$ to $2n_c$ on the laser axis (increasing exponentially to $10n_c$ at the cone wall interface), the laser is not able to reach the flat-top. Although the laser absorption has been increased to 83.2%, the maximum proton energy is decreased to 18.8 MeV due to the effect of the longer electron transport distance to the flat-top. Nevertheless, the simulated maximum proton energy is still more than 2 times higher than that from the simulated flat-foil target, an attribute of the enhanced laser absorption and hot electron temperature.

To gauge the effect of the longitudinal focusing, two simulations were performed such that the laser was focused both on the flat-top as well as 80 microns prior to the flat-top (the Rayleigh length is 100 microns in these conditions). The laser absorption, hot electron temperature and maximum proton energy are only slightly changed. This is consistent with the observation that the curved cone geometry dilutes the longitudinal pointing accuracy differences as seen in [Sentoku et al., 2003a]. The main reason for the increase in the maximum proton energy comes from the cone guiding the laser light allowing more energy transfer to electrons as well as a neck diameter being of the right dimensions so as to form a thick, underdense, preplasma for efficient hot-electron generation and a short undisturbed transport to the flat-top.

It has therefore been shown that laser-ion acceleration with flat-top cones is a novel and an efficient method to obtain high quality energetic ion beams. PIC simulations have also shown that the maximum energy of the accelerated ions is proportional to the hot electron temperature and density, and is inversely proportional to the hot electron transport distance. The cone wall smoothness is an important factor to optimize the maximum ion energy, while the longitudinal pointing accuracy has only a small influence on the final proton energy. As long as the laser axis is aligned with the cone axis and as long as the preplasma is not sufficiently dense to affect laser propagation toward the cone tip, the maximum ion energy depends only on the hot electron population characteristics, which is influenced by the cone geometry and not necessarily by the flat-top size. If the preplasma level is high or if the laser is not well aligned, even the higher electron temperatures and densities will not be sufficient to overcome the effect of the elongated electron transport path and associated scattering to the cone side-walls. The long transport results in proton spray emission from the cone side-walls and poor ion beam quality and efficiency.

New target concepts along with new ideas to achieve ignition of fusion targets with laser and particle beams are presently of high interest and have a wide range of applications in the field of high energy density physics. Studies of the cone target along with other shapes have paved the way to enable a better understanding of the cone physics allowing to develop the cone target presented in the rest of this section along with its relevance for an array of applications [Renard-Le Galloudec et al., 2010].

For such a target to give its full potential, because of the physics occurring in a cone, some criteria need to be met (the same as the ones above). The cone target needs to be precisely aligned. The laser enters the cone and starts hitting the faces when its diameter is about 3 to 4 times the inside tip size [Sentoku et al., 2004; Nakamura et al., 2007; Renard-Le Galloudec et al., 2008]. Under low preplasma conditions [Sentoku et al., 2004; Renard-Le Galloudec et al., 2008], so as to not destroy the conical shape the laser interacts with, the cone microfocuses the laser light into the tip [Sentoku et al., 2004]. At the same time, the laser interacts with the faces of the cone, creates electrons and guides them along the faces to the tip where the electron beam gets out [Renard-Le Galloudec et al., 2009]. This increases dramatically the electron density in the tip, enables a higher conversion efficiency of laser light into very energetic or hot electrons [Sentoku et al., 2004; Nakamura et al., 2007; Nakamura et al., 2004; Nakatsutsumi et al, 2007], and thus enhances both electrons and protons [Chen et al., 2005]. Note here that the cone, not the laser, defines the beam diameter [Renard-Le Galloudec et al., 2008]. A smaller cone angle produces more energetic electrons compared to a more open cone [Nakamura et al., 2007; Chen et al., 2005]. In addition, cones show an increased absorption of the laser light compared to flat targets [Nakamura et al., 2007], which makes them more efficient. More complex cone-based geometries have also been studied [Flippo et al., 2008] and also show an increased efficiency compared to flat targets.

In [Renard-Le Galloudec et al., 2010], both the inside and outside tip of the cone are slightly curved. Shaping the back of flat targets has been demonstrated to focus protons beams [Wilks et al., 2001; Ruhl et al., 2001; Roth et al., 2002; Patel et al., 2003; Snavely et al., 2007], it is however the first time that this concept is adapted to a cone geometry in order to use cones as an essential element of the particle beam production and reap the benefits of the increased efficiency of its shape. It does more than a standard flat or curved target. It adds three essential aspects. The first aspect is that making use of the cone faces by allowing the laser to spread on them greatly reduces the amount of preplasma filling the cone, thus enabling an efficient use of the cone shape. It also uses the faces to create the electrons and guide them to the tip. Several articles have showed the imprint of the laser pattern on flat targets into the particle beam [Fuchs et al., 2003]. As the laser bounces several times on the faces on its way to the tip, its imprint disappears. It creates, at the tip, a laser imprint free area of high energy density, enabling more uniform beams. Also, if best focus is positioned toward the entrance of the cone, then all of the laser light available gets in regardless of the fnumber of the focusing optic compared to the cone angle. The second aspect is the fact that the cone, then, not the laser, defines the particle beam [Renard-Le Galloudec et al., 2008]. The laser is clearly not directly driving the characteristics of the beam produced. The third aspect is the ability to control the divergence of the output beam. The tip of the cone is slightly curved in this case. This results in a modification of the divergence of the output particle beam by effectively modifying the accelerating sheath shape, and can be tuned by adjusting the amount of curvature. This efficiently produces a beam with extremely relevant characteristics to various applications. With high-energy high-repetition rate lasers as well as targets that are on the verge of cost effective mass production, cost effective compact applications can be readily envisioned.

Because the new target shape proposed in [Renard-Le Galloudec et al., 2010] has not been fabricated yet, the 2D Particle-In-Cell (PIC) code PICLS was used to run simulations and assess the electromagnetic fields structures and proton beam characteristics in comparison with flat targets. Several intensities were run to span the range available to short pulse lasers. The inner and outer tip diameters are respectively 10 and 30 μm. They are both curved. The target itself is 10μm thick. The incident laser pulse (1 μm, 40 fs, 21 μm FWHM transverse spot size at $3\times10^{18}$ W/cm²) has a Gaussian temporal and transverse spatial profile. The laser interacts with the target at normal incidence, with its polarization in the simulation plane. The initial target density is $40n_c$ and remains higher than the relativistic critical density $a_0n_c$, where $a_0$ is the normalized laser amplitude and $n_c$ is the critical density ($n_c = 1.1 \times 10^{21}/\lambda$ (μm)²cm⁻³, $\lambda$ is the laser wavelength). The plasma is composed of Al ions, protons and electrons.

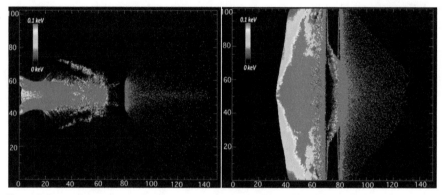

**Figure 6.** Proton energy density for the cone target at $3.10^{20}$W/cm² (6.a) and for the flat target for the same laser intensity (6.b). The pulse is injected to the left. Axis are in microns from 0 to 150 for the x-axis and 0 to 102 for the y-axis.

Figure 6.a represents the 2D proton energy density for a 10μm thick curved-tip cone in a high intensity case at $3.10^{20}$W/cm². Figure 6.b represents the same 2D proton energy density for a 10μm flat target at the same intensity. We clearly see that the protons are a lot more confined in the cone than in the flat target where they tend to diffuse laterally. The particles emitted from the cone are much more collinear to the laser axis compared to the flat target where they expand perpendicularly to the sheath.

In both cases the average divergence is small, especially for the high-energy protons. While the cone target does not seem to do better for these, it does control the divergence much better than the flat target over a wider range of energies. The curvature also allows to focus the most energetic protons in a specific location, and thus to deposit through the ions a higher energy in a smaller volume than in the case of a flat target, which is of special interest to isochoric heating.

The results in [Renard Le Galloudec et al., 2010] confirm that the cone is a much more efficient structure over a range of intensities (from $3\times10^{18}$ W/cm² to $3\times10^{20}$ W/cm²). As the intensity increases, the maximum proton energy increases in general regardless of the target but the cone target clearly shows higher maximum proton energy than the flat target for all intensities. That difference increases with intensity. Especially evident at $3.10^{20}$W/cm², both electrons (fig. 7.b) and protons [Renard Le Galloudec et al., 2010] are accelerated to higher energies in a higher number for the cone target. Enhanced laser interaction results in much higher maximum proton energies at high intensities (from 54.4 MeV to 98.7 MeV). Laser absorption is greatly increased in cone targets. In the high intensity case, laser intensity reaches a maximum of $2.4 \ 10^{21}$ W/cm² in the tip of the cone ($6.10^{20}$ W/cm² for the flat target), highlighting the microfocusing effect of the cone. In the low intensity case, the large preplasma present in both cases tends to give similar laser parameters evolutions, similar electric fields and a moderate increase of maximum proton energy (see Figure 7.a). In the high intensity case, the laser intensity and the longitudinal electric field reach significantly larger values in the cone leading to an important increase in maximum proton energy.

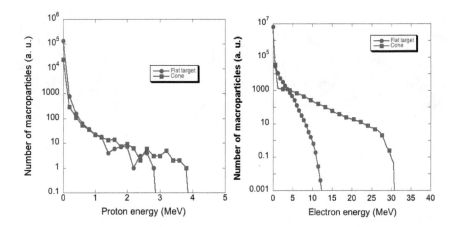

**Figure 7.** Proton energy spectrum (7.a) for both cone and flat target at 1.98ps for the low intensity case, and electron energy spectrum (7.b) for both cone and flat target at 924fs in the high intensity case.

It is therefore shown that this new conical target shape has the potential to produce proton beam of a higher maximum energy, a higher intensity and a lower divergence. Because of the appropriate use of the cone structure itself, by using the faces leading to the tip, it is also shown that the target itself defines the proton beams characteristics, nor the laser imprint or the focal spot size have an impact on these characteristics. The contrast of the laser can also be mitigated and finally the fnumber of the focusing optic is superseded by that of the cone target itself. All these parameters increase the potential for various groups to join in the research endeavour and pursue exciting new applications.

## 4.2. Laser ion acceleration using reduced-mass targets

The coupling between high intensity laser pulses with solid foils having a limited transverse extension (~ few tens of μm) has been studied by diagnosing electrons and protons produced during this interaction [Buffechoux et al. 2010]. It was observed that by reducing the area of the target surfaces, it is possible to reflect electrons near the edges of the target during and just after the laser irradiation. This transverse refluxing, which does not occur with usual large planar targets, can maintain a hotter, denser and more homogeneous accelerating electron sheath. As a consequence, when this transverse refluxing takes place during the ion acceleration duration, the maximum ion energy and the conversion of laser energy in the energy of high energy ions are strongly increased. The ion beam divergence is also reduced thanks to the more homogeneous electron sheath. These results will be detailed in this Section. Since, new experiments performed in 2011 using such targets allowed to obtain proton energies of around 80 MeV [Schollmeier et al., 2011].

The dynamics of MeV electrons generated in solids by ultra-intense lasers plays a crucial role in many applications such as electron-driven fast ignition [Tabak et al., 2006] or the production of secondary sources, e.g. X-rays [Quéré et al., 2006], positrons [Chen et al., 2009] or ions [Fuchs et al., 2009], all with important scientific or societal perspectives.

Understanding the hot electrons dynamics requires considering several aspects of their transport through the target. This Section reports an investigation of the influence of the target lateral dimensions on the dynamics of hot electrons and associated energetic proton production [Buffechoux et al., 2010]. It was the first to identify the important role played by the lateral electron recirculation in small targets. As observed in the simulations discussed here, electrons that are injected into the target center are seen to spread along the target surface with a velocity $v^t_{hot}$ ~0.7c and reflected at the target edges. They therefore transit from the center to the edges and back in a time $\tau_t = D_s/v^t_{hot}$ where $D_s$ is the target transverse diameter. When $\tau_t$ is of the order of the laser pulse duration $\tau_L$, the hot electrons, refluxing from the edges, are confined during or shortly after $\tau_L$. This leads to a time-averaged denser, hotter and more homogeneous electron population. This has been observed through the use of a combination of hot electrons and accelerated proton diagnostics.

The effective (time-averaged) hot electron temperature ($T_{hot}$) as a function of the target surface area was analyzed [Buffechoux et al., 2010]. It was observed that both $N_{hot}$ and $T_{hot}$ increase for targets having surface areas <3-4×10⁴ μm², corresponding to target transverse diameters $D_s$<170-200 μm. For larger targets, both $T_{hot}$ and $N_{hot}$ remain, on the contrary, constant. With regards to $T_{hot}$, the data are consistent with simulations of sharp-interface plasmas irradiated at 45° with S polarization [Lefebvre et al., 1997]. With regards to $N_{hot}$, the data are consistent with measurements of hot electrons density ($n_{hot}$) and sheath surface [Antici et al., 2008a] which, combined, yield comparable number of electrons contained in the sheath. All these results are consistent with previous complementary measurements that showed an increase of bulk target heating when reducing the target surface area [Nakatsutsumi et al., 2008; Perez et al., 2010].

It was also observed that the hot electron sheath becomes more uniform when the target surface area is reduced. This result is obtained by analyzing the angular proton beam profile as recorded on the RCF film. It is shown that the beam is more collimated when the target surface area is reduced. This suggests a flatter electron sheath along the target rear side for smaller targets, which is consistent with the picture of geometrically confined electrons. The measurement of the thermal emission from the target rear side further confirms this. As a result of the observed increase of $N_{hot}$ and $T_{hot}$ within the electron sheath, when reducing the target surface area, a clear improvement of the proton beam characteristics was observed (maximum proton energy and laser-to-proton conversion efficiency).

Two-dimensional (2D) particle-in-cell (PIC) simulations of laser target interactions were performed to help identify that lateral refluxing of the hot electrons is the key process leading to hotter, denser and more homogeneous electron sheaths when reducing the target surface area. The mechanism of refluxing, as observed in the simulations, is as follows: because of their high velocity, electrons trajectories can be considered as ballistic. As electrons enter the target with an angle close to the laser incident angle (45°), their average transverse (with respect to the target normal) velocity in the target is $v^{t}_{hot} \approx c \times \cos(45) = 0.7c$. After several reflections off the sheath fields on the front and back surfaces with that same angle, the electrons will have travelled transversely to the edge of the target to be again reflected back (note that here there is no distinction between electrons turning around the target and electrons reflected back).

Since the electric field accelerating the ions is proportional to $(n_{hot}T_{hot})^{1/2}$ [Fuchs et al., 2009; Mora, 2003], proton acceleration is expected to be enhanced for the small foil compared to the medium foil, and indeed the simulation results agree well with the variation of the experimental proton cutoff energies. The enhancement in ion acceleration can take place only if the electrons can come back to the target center within the ion acceleration time $\tau_{acc}$. From this, an experimental proton acceleration time can be deduced $\tau_{acc} \approx 800\text{-}950$ fs. This is consistent with theoretical estimates. Also in good agreement with the experimental result, simulation results show that refluxing produces in the simulations a more uniform sheath in small targets. The constrained lateral dimension forces the hot electrons to recirculate in the small foil, thus homogenizing the sheath.

Reducing the surface area of solid targets therefore leads to an increase in the effective hot electron number and mean energy due to the lateral electron recirculation. In particular, this effect enhances the properties of laser-accelerated ions, in terms of energy, flux and collimation, as necessary for progress towards applications. Use of this simple mechanism is of interest for future experiments at higher intensity laser facilities, but will impose high temporal contrast laser pulses to avoid preplasma leakage to the target rear-surface. In addition, progress in target fabrication will offer targets that are not only of reduced lateral size, but also thinner as it is now found favorable for ion acceleration (see Section 2.3).

## 4.3. Focalization of laser accelerated ions using curved targets

The dynamics of the focusing of laser accelerated ions using curved solid targets has been intensively studied in the last years [Patel et al., 2004; Kar et al., 2011; Chen et al., 2012]. The ability to tightly focus, i.e. over tens of microns, dense (> $10^{10}$ particles), short (with duration ~ps) bunches of positively charged particles is far beyond the possibility of present particle accelerators. The increase of the particle density of these beams would significantly improve the efficiency and prospects of a number of important applications. High energy density proton beams would allow, for example, igniting pre-assembled Inertial Confinement Fusion (ICF) targets, also known as proton Fast Ignition (PFI) [M. Roth et al., 2001], or enable ultrafast heating, above keV, of dense materials to explore their properties [Mancic et al., 2010; Pelka et al., 2010; Carrié et al., 2011]. High energy density positron beams would allow enhancing the rate of antimatter creation in the laboratory [Andresen et al., 2007], opening up investigations of many fundamental laws of nature.

These positively charged beams, broadband or monoenergetic, can nowadays be produced with a high particle number and over a short duration using ultra-intense, short pulse lasers interacting with solids targets [Wilks et al., 2001; Chen et al., 2010] as described in Section 2. When using flat targets as sources, the ion beam is divergent, 0-25 degrees depending on proton energy [Fuchs et al., 2003; Roth et al., 2005; Toncian et al., 2006], since the expanding sheath field front on the target rear-side is Gaussian in shape [Romagnani et al, 2005; Antici et al., 2008]. Such divergence can however be compensated by curving the back surface of the target, so that the accelerated proton beam will converge [Patel et al., 2003; Kar et al., 2011; Offermann et al., 2011]. This points to the necessity of understanding and optimizing the dynamics of the focusing sheath in order to achieve tight focusing, as necessary for the above-mentioned applications.

A study of the ion beam focusing dynamics through temporally and spatially resolved measurements, with picosecond and micrometer resolution, of the shape of the sheath field from a curved target, irradiated by a high-intensity short pulse laser has recently been performed [Chen et al., 2012]. Experiments performed at LULI show that the major part of the energy carried by ions converge at the center of cylindrical targets in a spot having a diameter of 30 microns, which can be beneficial for applications requiring high ion energy densities. The location of the focus is a function of proton energy, although most of the protons focus at the geometric target center. It was also shown experimentally and using 2D PIC simulations that the exact location of laser illumination on the curved target does not adversely affect the ability to focus the sheath-accelerated ion beam, although it modifies the directionality of the ion beam. Despite these advantages, an important filamentation was observed during the focusing, which limits the energy deposition precision in the ion converging zone. This effect is important for the use of such a setup in order to obtain high ion energy densities. It is the case in the fast ignition concept with ions which requires to have a high ion concentration zone with a diameter of 10 to 20 microns. Simulations have shown that at higher laser intensities, closer to the ones expected to be used for ion fast ignition, ion focusing was improved.

## 4.4. Acceleration of quasi-mono-energetic ions beams using the TNSA mechanism

For many potential applications of laser accelerated ions, controlling the characteristics of the produced ion beam is essential. Several techniques have recently been tested to obtain quasi-mono-energetic ion beams using the TNSA mechanism. Using either a secondary laser to control the thickness of the layer of ions accelerated to high energies [Hegelish et al., 2004], or small plastic dots to reduce the transverse extension of the accelerated proton bunch [Schoewer et al., 2004], these techniques try to constrain the acceleration zone to avoid acceleration gradients and therefore limit the ion energy spread. Another possibility to control the energy spread as well as the beam divergence has recently been developed. This technique consists in using a tuneable plasma microlens that allows focusing the ion beam and selecting specific energies. The control of the characteristics of proton beams using a solid cylinder irradiated by a secondary laser has therefore been studied [Toncian et al., 2006; d'Humières et al., 2007] using simulations and experiments performed at LULI (France). This Section explains these various techniques and discusses their limits (Figure 8).

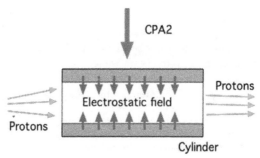

**Figure 8.** Schematic of the micro-lens setup focalizing a proton beam propagating on its axis. The proton beam is accelerated from a planar metallic foil by a first laser pulse. The proton beam is focalized using the hollow cylinder irradiated on the side by the laser pulse CPA2.

The focalization is effective only for a range of energies: as the cylinder is located at a certain distance from the source, and as the source has a large spectrum, protons with different energies have different times of flight and therefore reach the cylinder at different times. If the high energy protons cross the cylinder before it is irradiated by the secondary laser generating the focalizing fields, they do not undergo any focalizing effect (it is the case of the picture on the left in Figure 9). Protons with a slightly lower energy, synchronized with the focalizing fields will undergo the maximum effect. This is illustrated by the simulations performed in Figure 9.

The focalization symmetry is also observed in PIC simulations [d'Humières et al., 2006] and comes from the fact that fast electrons, even if they are produced in one region of the cylinder wall, propagate faster on the whole diameter than it takes them time to expand in vacuum towards the interior of the cylinder. Therefore, when the plasma expansion into vacuum phase, at the origin of the focalizing radial electric fields, starts, the radial distribution of these fields is symmetric.

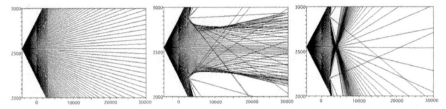

**Figure 9.** Trajectories of 100 protons at (a) 7.6 MeV; they exit the cylinder (located at x=0) before it is irradiated and their divergence is not affected. The two axis are in microns but at different scales. (b) Protons at 6.25 MeV; they are close to the exit of the cylinder when it is irradiated, sustain small fields and are therefore well collimated. (c) Protons at 4.9 MeV; they are at the middle of the cylinder when it is irradiated, they therefore sustain fields much stronger than the ones on b). They are therefore focalized at a short distance (5 mm) from the cylinder and then strongly diverge.

It appears that this micro-lens (its diameter is typically of the order of a mm, as well as its length) has the advantage, compared to curved targets, to allow a focalization which is not limited to distances of a few mm. It also allows to move away the source from the focalization region, which can be interesting for instance in the perspective of fast ignition using proton beams: increasing the distance between the two would allow to avoid damaging the source during the implosion of the main target. The focalization distance can indeed be tuned simply by varying the intensity of the beam generating the focalizing fields.

The fact that the focalization distance is variable as a function of energy can be exploited to select in energy in a controllable manner part of the spectrum, which is initially large. It is then enough to position a small opening at a calculated distance from the cylinder. A small number of high energy protons (crossing the cylinder before the focalization) will go through the opening. The protons with a focalizing distance corresponding to the distance between the opening and the cylinder will all be transmitted leading to a peak in the spectrum while the protons with lower energies will be focalized before the opening and will therefore reach it unfocused, and a small number of them will cross the opening. The position and the width of the peak can be easily adjusted by tuning the intensity and the delay between the pulse triggering the micro-lens and the primary pulse generating the source, as well as by tuning the distance between the micro-lens and the source.

The use of the micro-lens is not limited to protons. Therefore, the micro-lens presents the advantage, in comparison with previous solutions, of being able to achieve focalizing as well as energy selection in a single step. Moreover the process is simple: the pulse triggering the micro-lens requires just 10% of the main pulse, the micro-lens is cheap (it is just a section of what is used for medical syringes), it does not require any particular target engineering, and finally it is easy to change the desired parameters (focalization distance, size of the focalization point, energy of the protons to be focalized, peak to be selected in energy, width of this peak) by adjusting the laser parameters (intensity, delay between the two pulses) and the cylinder-source distance.

This development, achieving compact collimation of energetic ions, is therefore an extremely important step towards applications of laser accelerated ion sources. These advantages compared to existing (conventional) setups can be summarized as:

- Compactness: mm instead of dm.
- Capability to focus beams of several A instead of several mA; the last two points are the results of the fact that it is a plasma setup, therefore able to sustain much higher electric fields than conventional setups.
- Selective collimation in a few ps.
- Capability to focus in a few mm, cm or m multi-MeV beams in place of a minimum of a few meters for conventional setups or not more than mm for curved targets.

## 5. Laser ion acceleration limits in the ultra high intensity and ultra high laser energy regimes

With the recent rapid progresses in high power laser technologies, new laser systems are under construction and will allow to explore the ultra high intensity regime (Apollon in France and then ELI in Europe) and the ultra high energy regime (OmegaEP at Rochester, PETAL in France, ARC at LLNL and FIREX in Japan). This section presents the limits of the laser ion acceleration mechanisms described in the previous sections in these extreme regimes. These studies aim at obtaining a better understanding of how the predictions given by the existing laser ion acceleration models will be affected in these regimes and to estimate more realistically their potential.

### 5.1. Laser ion acceleration in the ultra high intensity (UHI) regime: Effects of radiation losses

Several PIC codes have recently been modified to take into account radiation losses by charged particles [Sokolov et al., 2009a; Tamburini et al., 2010; Capdessus et al., 2012] These models were then applied to the study of laser ion acceleration in the ultra high intensity regime (> $10^{22}$ W/cm$^2$) [Naumova et al., 2009; Tamburini et al., 2010; Capdessus et al., 2012; d'Humières et al., 2012] when radiation losses become important and can not be neglected anymore. This section discusses these results and shows how, in this regime, radiation losses strongly depend on the target density, thickness and on the laser pulse polarization. Even if radiation losses always lead to electron cooling, their effect on the ion distribution depends on the interaction conditions. For thin and moderate density foils, radiation losses can even enhance laser ion acceleration while decreasing electron heating [Capdessus et al., 2012]. These results have been confirmed using 2D PIC simulations [d'Humières et al., 2012].

A new generation of laser systems such as planned in the Extreme Light Infrastructure (ELI) project [http://www.extreme-light-infrastructure.eu/] will produce laser intensities as high as $10^{24}$W/cm$^2$. New physical processes are expected under these conditions such as emission of high energy photons, the radiation reaction force acting on electrons, electron-positron pair production, acceleration of ions to relativistic energies, etc. [Bulanov, 2009]. One of the important applications of ultra intense laser pulses is acceleration of charged particles to extremely high energies. Recent numerical simulations and theoretical analysis show that at laser intensities exceeding $10^{22}$ W/cm$^2$ the ions can be accelerated to relativistic energies under the laser radiation pressure. While the radiation reaction force has been applied to the motion of a single particle for a long time, it has only recently been considered in plasma physics. First kinetic simulations of laser plasma interaction with a particle-in-cell (PIC) code accounting for the radiation force were reported in Refs. [Naumova et al., 2009; Schlegel et al., 2009; Tamburini et al., 2010, Chen et al., 2011]. They demonstrated the role that the radiation reaction plays in the radiation pressure acceleration by high intensity laser pulses.

The effect of radiation losses on the process of ion acceleration by ultra intense laser pulse has also been studied in 1D in [Capdessus et al., 2012]. This effect becomes important for laser intensities exceeding $10^{22}$ W/cm$^2$, where the radiation friction force slows down electrons and affects the ion dynamics through the self-consistent electrostatic field. The effect of radiation losses depends strongly on the target density and thickness and also on the laser polarization. It is less important in the case of strongly overdense targets and for a circular polarization, where the relative density $n_e/n_c$ is larger than the laser amplitude, $a_L$. This is explained by clear spatial separation of the particles and fields. On the contrary, the radiation losses are important in the induced transparency regime where $n_e/n_c < a_L$. Although radiation losses are always leading to cooling of electrons, their effect on the ion distribution depends on the target thickness. In the case of thin targets, where the areal density is small, radiation losses may improve ion acceleration. On the contrary, in the piston regime, radiation losses lead to a reduction of the piston velocity and less efficient ion acceleration. These simulation results were limited to 1D simulations. However, the particle momentum has three components and arbitrary angles of electron propagation with respect to the laser wave are accounted for. No qualitative changes in 2D or 3D simulations are therefore expected. Recent results published in [Tamburini et al., 2012] confirm that statement. They demonstrated that results obtained for lower dimensionality remain valid qualitatively, although the maximum energy of ions in 3D is found to be higher than in corresponding simulations in 1D and 2D. The self-generated magnetic fields are also expected to be responsible for stronger radiation emission.

2D simulations were performed in [d'Humières et al., 2012]. The goal was to study how laser ion acceleration using thin overdense targets evolves in the ultra high intensity regime (> $10^{22}$ W/cm$^2$) and to assess the importance of radiation losses in this regime. The radiation losses model used is based on the so-called Sokolov model [Sokolov et al., 2009a]. In a first set of simulations, the wavelength of the incident pulse is 0.8 μm, its pulse duration is 21 fs and its irradiance is $1.6 \times 10^{22}$ W/cm$^2$. The FWHM of the focal spot is 5 μm. The p-polarized Gaussian pulse interacts with the target in normal incidence. The plasma is composed of protons and electrons with a constant 400 $n_c$ density. The plasma slab thickness was varied from 0.2 to 0.8 μm. Figure 10.a shows the simulated proton spectra with and without radiation losses for the 0.5 μm case. In this range of thicknesses, radiation losses have a measurable effect on ion acceleration but this effect remains small. In a second set of simulations, the wavelength of the incident pulse is 0.8 μm, its pulse duration is 48 fs and its irradiance is $1.5\times10^{23}$ W/cm$^2$. The FWHM of the focal spot is 5 μm. The circularly-polarized trapezoidal pulse interacts with the target in normal incidence. The plasma is composed of deuterons and electrons with a constant 10 $n_c$ density. The plasma slab thickness was varied from 0.4 to 5 μm. Figure 10.b shows the simulated electron spectra with and without radiation losses for the 5 μm case. For such a high intensity the radiation losses effects become important and the energetic electron population is strongly affected. Very high ion energy are nevertheless obtained (a maximum energy of 1.8 GeV for Deuterons in the 5 μm case) [d'Humières et al., 2012].

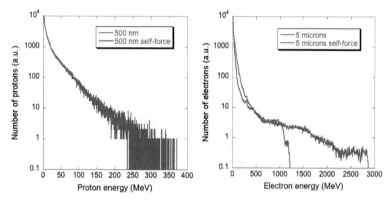

**Figure 10.** (a) Proton spectra for the $1.6\times10^{22}$ W/cm$^2$ pulse interacting on a 500 nm high density foil. (b) Electron spectra for the $1.5\times10^{23}$ W/cm$^2$ pulse interacting on a 5 μm overdense foil. Blue curve: without radiation losses. Red curve: with radiation losses.

The interaction of a high intensity short pulse with overdense targets is now actively studied in the UHI regimes to estimate ion acceleration possibilities on future laser facilities. This study is performed using Particle-In-Cell codes in which charged particles energy losses through radiation have been implemented. In the ultra high intensity regime, radiation losses will start affecting laser ion acceleration using thin overdense targets for intensities higher than $10^{22}$ W/cm$^2$, but maximum proton energies of a few hundreds MeV can still be reached.

## 5.2. Laser ion acceleration in the ultra high energy (UHE) regime

The interaction of a ultra high energy (up to 3.5 kJ) and high intensity laser with various types of targets has recently been modelled using PIC simulations to prepare laser ion acceleration experiments on high energy installations [d'Humières et al., 2012]. Laser ion acceleration in the ultra high energy regime (> 1 kJ) has also recently been studied experimentally [Flippo et al., 2010]. This section presents a study of this regime to analyze the changes brought by such high laser energies on the main accelerating mechanisms presented in the previous sections. Energies of more than 100 MeV can be obtained in a robust manner using solid foils with thicknesses of a few tens of μm. These beams could then be used to radiograph the implosion of a DT target when coupling PETAL with the LMJ for instance, or for other experiments (isochoric heating, laboratory astrophysics…).

A new era of plasma science started with the first experiments on the National Ignition Facility (NIF) at the Lawrence Livermore National Laboratory (LLNL) in the USA. The Laser MégaJoule (LMJ) under construction near Bordeaux in France is following the trail opened by the NIF with its planned 160 laser beams for more than 1 MJ to reach ignition of a deuterium - tritium target using the indirect drive method. Besides the physics of ICF (plasma physics, shock / fast ignition), NIF & LMJ will be essential for basic science, exploring fields such as plasma astrophysics (e.g. study of shocks to simulate violent events in the Universe such as supernovæ, accretion disks), planetary physics (highly compressed and warm matter), stellar interiors with large coupling between radiation field and matter & nuclear physics. A petawatt short pulse laser will be added to the ns pulse beams of the LMJ. This is the PETAL system (PETawatt Aquitaine Laser) [Blanchot et al., 2008], under construction on the LMJ site near Bordeaux (France) and funded by the Région Aquitaine. The ultimate goal is to reach 7 PW (3.5 kJ with 0.5 ps pulses). For the beginning of operation, the PETAL energy will be at the 1 kJ level, corresponding to an intensity on target of $\sim 10^{20}$ W/cm$^2$.

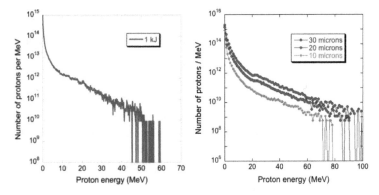

**Figure 11.** (a) Computed proton energy spectrum emitted from the PETAL target with 1kJ energy. (b) Variation of the proton energy spectra for three laser FWHM: 10 μm (green line), 20 μm (red line) and 30 μm (blue line) with 1.75 kJ at 2ω.

In the experiments proposed for LMJ - PETAL, the petawatt laser may be focused on a secondary target, where a short (~ 20 ps length) bunch of particles (electrons, protons, ions) is produced and may be used to probe the plasma generated by the ns LMJ pulses. The energetic ions are produced using the TNSA mechanism detailed in Section 2. First calculations were performed using the 2D particle-in-cell code PICLS to compute the energy spectra and angular divergences of the protons produced with PETAL (Fig. 11.a) for 1 kJ of energy in the PETAL beam and for 3.5 kJ. Protons with a maximum energy higher than 100 MeV are measured in the 3.5 kJ case. Within these conditions, these calculations can be considered as majoring the intensities and electron energy spectra end-points. After the accurate characterization of the particle emission from the PETAL target, the diagnostics will be used for plasma experiments, which, for example, requires proton radiography to determine the magnetic field or the electric field structure at the plasma scale or to measure the density of the LMJ plasmas.

Multi-kJ Petawatt-class laser systems therefore open new and exciting opportunities for laser ion acceleration. The interaction of a high intensity short pulse with underdense, near-critical and overdense targets has been studied using 2D Particle-In-Cell simulations in these regimes [d'Humières et al., 2012]. The goal was to study laser ion acceleration in the ultra high energy regime (> 1 kJ). 2D Particle-In-Cell simulations were performed to study the interaction of the expected PETAL laser pulse with a thin target in the transparency regime [Dong et al., 2003; d'Humières et al., 2005]. The wavelength of the incident pulse is 1 μm, its pulse duration is 500 fs and its irradiance is $10^{21}$ W/cm$^2$. The FWHM of the focal spot is 30 μm. The p-polarized Gaussian pulse interacts with the target in normal incidence. The plasma is composed of protons and electrons with a constant 400 $n_c$ density. The plasma slab thickness was varied from 0.2 to 1 μm.

The maximum proton energy measured in the simulation depends on target thickness and reaches several hundreds of MeV at 1ω. The results in [d'Humières et al., 2012] indicate that the most energetic protons come from the rear surface of the target. It will not be straightforward to accelerate protons with such thin targets without controlling the laser contrast. A plasma mirror or doubling the frequency could improve this contrast. With a 250 nm foil and at 2ω (limiting the total energy to 1.75 kJ), the maximum proton energy is lowered to 160 MeV and $3.1\times10^{10}$ protons/MeV at 100 MeV are expected. Figure 11.b shows the variation of the proton energy spectra for three laser FWHM: 10 μm (green line), 20 μm (red line) and 30 μm (blue line). As expected, the number of protons at high energies increase with the laser FWHM. The maximum proton energy also increases but at a slower rate. A simulation using a 10 microns CH target at solid density with a preplasma and similar laser parameters predicts a maximum proton energy of around 110 MeV.

Another promising way to accelerate ions to high energies is to use underdense or near-critical density targets (see Section 3). The interaction of a laser pulse with a near-critical density target and an underdense target was simulated using the above laser parameters in 2D. In both cases the plasma is composed of protons and electrons. For the first case, the target density is constant and its thickness is 100 μm. For the underdense case, the target density profile is a cosine square with a full-width at half maximum of 100 μm and a maximum density of 0.4 $n_c$. A high laser absorption of 75.9% is measured in the 2 $n_c$ case and 18.9% is measured in the 0.4 $n_c$ case as the laser propagates through the plasma leaving a large channel behind. The generated hot electron population produces strong electrostatic fields at the back of the target.

Figure 12 shows the proton spectra in both cases (a) and the proton phase space at the middle of the simulation box on the laser propagation axis 0.99 ps after the beginning of the simulation (b). For both cases the most energetic protons come from a region close to the back surface of the target. Wave breaking is clearly visible in the underdense case as expected from previous theoretical studies. The maximum proton energy in the near-critical case is more than 400 MeV, which is comparable with the solid density target simulation, and $7 \times 10^9$ protons/MeV at 200 MeV are expected. It reaches a little more than 200 MeV in the underdense case with $2 \times 10^{10}$ protons/MeV at 100 MeV.

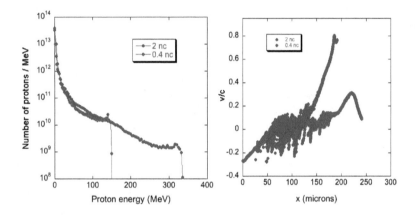

**Figure 12.** (a) Proton spectra for the 2 $n_c$ case (blue line) and for the 0.4 $n_c$ case (red line). (b) Proton phase space at the middle of the simulation box on the laser propagation axis 0.99 ps after the beginning of the simulation.

Interaction conditions are not optimized. In this low density regime it is possible to find an optimum thickness and a key parameter controlling the acceleration process is the product of the density and the thickness. This optimum thickness also depends on laser parameters. The interaction of a PETAL laser pulse with a low density target could therefore lead to even higher maximum proton energies than solid targets. The near-critical target could be achieved experimentally using a foam, an aerogel or a cryogenic target, the underdense target using a gas jet or a foil exploded by a nanosecond pulse (see Section 3).

The interaction of a high intensity short pulse with underdense, near-critical and overdense targets has therefore been studied with 2D PIC simulations in the UHE regime to estimate ion acceleration possibilities on future laser facilities. In the ultra high energy regime, proton beams with maximum energies of several hundreds of MeV and a high number of high energy protons could be accelerated using thin solid foils or underdense and near-critical targets.

# 6. Applications of laser ion acceleration

As described in the previous sections, laser ion acceleration has already been applied to diagnose electromagnetic fields in plasmas and to produce warm dense matter. Other important applications are envisioned in laboratory astrophysics, as injectors for conventional accelerators or for the production of radioisotopes and hadrontherapy. Several of these promising applications of laser accelerated ions are presented in this section. This list, which is not exhaustive, aims at illustrating the important potential of compact high intensity ion accelerators.

## 6.1. Laser accelerated ions as a diagnostic of hot electron transport in dense targets

The possibility to control the shape of MA electron currents in solids by high intensity laser pulses has been studied using different target materials [Sentoku et al., 2011]. This Section details how these effects can be diagnosed using laser accelerated ion beams as the accelerating electron sheath is affected by the transport of the hot electron beam. These results are beneficial to applications like the production of secondary sources and the ion fast ignition concept for inertial confinement fusion.

Ultra-high currents (MA) of suprathermal (MeV) electrons, that are driven through solids using relativistic laser pulses (with intensity $I > 10^{18}$ W/cm$^2$), lie at the heart of numerous applications, either present-day like the generation of ultra-short secondary sources of particle and radiation (ions [Malka, 2008], X-rays [Mancic et al., 2010b], positrons [Chen et al., 2010], or neutrons [Disdier et al., 1999]), or potential like fast ignition of inertial confinement targets [Tabak et al., 1994] or laser-driven hadrontherapy [Bulanov et al., 2002a; Fourkal et al., 2002; Malka et al., 2004]. Having the possibility to spatially shape these currents is crucial in order to optimize the efficiency of each process. Sentoku et al. [2011] showed by combining experiments and modelling, how dynamical shaping of the currents can be achieved in various conductor materials. By tuning the target ionization dynamics that depends both on the target material properties and on the input electron beam characteristics, it is possible to control the growth of resistive magnetic fields that feedback on the current transport. As a result, collimating, hollowing or filamenting the electron beam can all be obtained and observed on the laser accelerated ions characteristics.

For all these applications, metals are preferable as target materials as they can provide enough cold electron return current to neutralize the forward laser-generated current and allow its propagation at current levels (MA) exceeding the Alfven critical current [Alfven, 1939]. At solid density, such targets are little prone to the Weibel relativistic electromagnetic two-stream instability [Weibel, 1959; Sentoku et al., 2003b] that is collisionally damped [Sentoku and Kemp, 2008]. Instead, resistive magnetic fields have been suggested to play more of a role [Kar et al., 2009; Robinson et al., 2007; Storm et al., 2009; Solodov et al., 2010]. These fields, that could collimate or focus the otherwise divergent electron beams [Debayle et al, 2010; Perez et al., 2010b], are driven by the Ohmic fields with the resistivity dynamically changing during target heating due to the hot electrons flow. The fields evolve according to Faraday's law. Weighting the role of magnetic fields has been relying up to now on hybrid simulations. These however present significant limitations due to their inability to self-consistently model the laser-generated hot electron source evolving at the dynamically ionizing interface. This has thus limited so far our understanding of the magnetic fields influence and impacted the possibility to adequately and quantitatively plan future progress, e.g. in designing fast ignition targets [A.A. Solodov et al., 2009] for present large-scale international programs [M. Schirber et al., 2005]. It is now possible to use Particle-In-Cell codes treating relativistic collisions and ionization processes to solve this issue.

Modulations of the sheath potential near the target surface will be imprinted in the protons angular distributions as they are detaching from the rear surface during the acceleration [H. Ruhl et al., 2004]. In [Sentoku et al., 2011], the 10 μm thick Au target exhibits a single peak distribution with the tightest profile, consistent with the experimental observation. The 15 μm thick Cu target on the other hand has a twin peak distribution, also consistent with the doughnut pattern of the experimental measure. Finally, the 40 μm thick Al target has a lower potential, since it is the thickest target, with a wider and modulated distribution, again consistent with the experiments. The trends observed in the experiment when increasing the target thickness or reducing the laser energy, i.e. a disappearance of the modulation in Cu or Al, are also observed in simulations performed with PICLS using collisions and ionization. For a 40 μm Cu target, strong ionization ($Z>15$) proceeds in a distance $x \sim 15$ μm, which is consistent with the heated region seen in the electron energy density map. Thus, strong resistive magnetic fields only grow in this region before breaking into weak filaments, $\sim 5$ MG, that cannot modulate strongly MeV electrons. As a result, the electrons spray out and form a smooth sheath potential at the target rear surface. Similar simulation results from reduced laser energy cases were also obtained.

The technique of manipulating the target resistivity developed here allows, even using a monolithic material, to control MA current flows in solids, e.g. excite the pinching, hollowing or filamenting of the currents. For this, the target thickness should be equal to or a bit thinner than the propagation distance of ionization waves during the laser pulse duration. Beyond the demonstration in [Sentoku et al., 2011], an important practical point is that keeping the same laser energy but changing the laser intensity (by e.g. defocusing the laser or increasing the pulse length), one can engineer transporting the same electron charge through different spatial form, e.g. pinched or hollowed out. Diagnosed using laser accelerated protons, it could help get higher maximum proton energies and smoother proton spatial distributions.

## 6.2. Magnetic lens for laser accelerated ions

High resolution proton deflectometry has been applied to measure magnetic fields in dense plasmas in order to quantitatively compare with simulations of the evolution of the magnetic field produced by intense electron currents generated by high intensity short laser pulses [Albertazzi et al., 2012]. Indeed, as seen in Section 6.1, such magnetic fields have the potential to limit the divergence of fast electron beams, a crucial factor in the context of secondary sources production or for achieving fast ignition of fusion targets. The capacity to measure these fields allows to have a better understanding of the implied physical mechanisms and to optimize them. In Section 6.1, these fields were diagnosed directly by ions leaving the target. In this section, two targets are used: one to accelerate protons and a second one irradiated by a secondary laser and radiographed by the proton beam coming from the first target.

Preliminary studies [Sentoku et al., 2003b] have shown that one of the main mechanisms to achieve collimation of of an energetic electron beam could rely on self-generated magnetic fields. These fields grow in the solid since the fast electron beam is not neutralized everywhere by the cold return current. The magnetic field arises from the resistive electric field $\mathbf{E} = \eta \mathbf{j}$ which depends on the resistivity $\eta$ and on the current $\mathbf{j}$. Taking into account Maxwell-Faraday equation, two terms for the temporal derivative of the variation of the magnetic field can be obtained [Sentoku et al., 2011]. One term is a current term and the other one is a resistive term. This last term depends on the collision time, atomic number of the target, temperature and time since the resistivity will evolve in time. In order to study the influence of the self-generated magnetic fields on the transport of fast electrons, the self-generated magnetic fields were measured using face-on proton radiography and the influence of target material was also studied [Albertazzi et al., 2012]. Using a plasma mirror and thin targets, it was possible to probe self-generated magnetic fields in dense target. The experiment gives the dynamics of this self-generated magnetic field over a very large time: ~100 ps. The temporal dependence of the magnetic field patterns shows a field inversion which appears at different times for different materials. 2D PIC simulations including ionization and collisions performed with the code PICLS and coupled to test particles simulations using the fields obtained with the PIC simulations have allowed to reproduce some important features of the experiments and thus to better understand the physical mechanisms at play. This setup can also be used to constitute a magnetic lens to improve the characteristics of laser accelerated ions (divergence and energy spread).

## 6.3. Collision of laser accelerated plasmas to study collisionless shocks for astrophysics

Ion beam instabilities play an essential role in the formation of collisionless shocks in laser-plasma interaction or in astrophysics. Weakly relativistic shocks are considered as potential sources of high energy cosmic rays. Experiments performed in the laboratory would allow to have a better understanding of the microphysics involved in this process. As demonstrated in the previous sections, laser systems delivering short high intensity laser pulses can accelerate ions to high energies. Low energy electrons neutralize these ion beams. It is therefore possible to produce counter-propagating beams to study the collision of fast plasmas and the formation of collisionless shocks (Figure 13) [Kato et al., 2008; Davis et al., 2010; Davis et al., 2012].

Collisionless shocks are an ubiquitous phenomena in the Universe. Weakly relativistic shocks are created in supernova explosions. It is believed that relativistic shocks are responsible for strong electron heating, magnetic field generation and subsequent emission of hard X and gamma rays in Gamma Ray Bursts (GRB). Collisionless shocks are considered to be a source of energy redistribution in Nature and high energy cosmic rays. However, a detailed mechanism of energy transformation of fast plasma flows into relativistic electrons and large amplitude magnetic fields is not known. The hypothesis of energy equipartition in a collisionless shock between the electrons, ions and magnetic fields has been shown in recent astrophysical models using large scale numerical simulations [A. Spitkovsky, 2008]. Laboratory experiments may provide further understanding of this obscure process. However, acceleration of sufficiently large volumes of matter to relativistic energies requires concentration of high densities of energy in a short time scale. Only high energy short pulse laser systems could be suitable for such experiments. The Target Normal Sheath Acceleration (TNSA, see Section 2) [Snavely et al., 2000; Wilks et al., 2001] can provide protons with energies of several tens of MeV corresponding to the streaming velocities of 20 – 30% of the velocity of light. The number of fast ions produced in one shot exceeds $10^{12}$, which should be sufficient for the collective processes to become significant. The range of such protons is of the order of 1 g/cm$^2$, so the collisional effects should not be significant for distances of many ion inertia length, which define the characteristic thickness of the shock front.

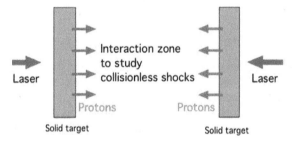

**Figure 13.** Schematic of the experimental setup required to perform the experiments discussed in this Section.

It was theoretically predicted that collisionless shocks may occur at mildly relativistic streaming velocities with a fraction ~ $10^{-4}$ of the total energy converted into magnetic field unless the electrons are heated significantly [Kato et al., 2008]. Experiments with non-relativistic laser pulses of sub-nanosecond duration and energy of a few hundred joules demonstrate a possibility of plasma streams formation with high Mach numbers, but the plasma temperature is relatively low and the ion mean free path remains shorter or comparable to the shock front thickness [Kuramitsu et al., 2011; Davis et al., 2010]. In a previous paper [S. Davis et al., 2010] it was proposed to collide the TNSA accelerated ion bunch with appropriately chosen low density plasma so the collective effects will dominate the ion stopping. [S. Davis et al. 2012] is dedicated to a more detailed study of electron heating and magnetic field generation in the overlapping region of the colliding plasmas: the first stage of formation of a collisionless shock. The characteristics of the high energy proton bunch were optimized in large scale PIC simulations. The ions were accelerated from an overdense target to sub-relativistic velocities by the TNSA mechanism. The plasma instabilities are excited by the streaming ions. The overall interaction time of about one ion plasma period, $\omega_{pi} t_{max} \sim 6$, is not sufficient for excitation of ion instabilities and full shock formation. In this time interval one can see the development of electron instabilities, magnetic field generation and electron heating.

This study has allowed to show that during the beginning of the plasma collision, a strong electron heating is generated by a Weibel type filamentation and each ions beam is then slowed. The energy partition between ions, electrons and electromagnetic fields, at the foundation of astrophysical models of gamma-ray bursts was therefore studied and has allowed to have a better understanding of the mechanisms at play. The better knowledge of laser ion acceleration described in the other sections of this chapter allow to envision collisions with even higher velocities with more energetic lasers (see Section 5.2). This would allow to get closer to relativistic shocks and study other types of astrophysical shocks (pulsar winds, other types of shocks involved in gamma-ray bursts...). Using an external magnetic field, it will be possible to study shocks with different magnetizations (ratio of the magnetic energy density and of the kinetic energy density) and thus to study in the lab for the first time collisionless shocks with parameters similar to the ones in supernovae remnants (SNRs).

## 7. Conclusions and perspectives

Laser acceleration of intense, collimated and multi-MeV ion beams is a recent research domain, rising fast. Its advent has been made possible thanks to short pulse lasers with extremely high intensities and progresses at the fast pace of the development of these lasers. Because of its pulsed nature (the duration of the source is of the order of a few ps), its high beam quality, of the high number of produced ions ($10^{11}$-$10^{13}$), of the possibility to modulate the spectrum as well as the divergence of the beam, the ion source produced by laser appears useful and promising for a number of applications. One can quote high temporal and spatial resolution radiography, fast ignition, production of warm dense matter and, later on, high intensity injectors for accelerators and sources for protontherapy or radio-isotope production.

This chapter presents several important contributions to the domain of laser ion acceleration and its applications. These contributions, experimental, theoretical and numerical, cover the possibilities to obtain in various regimes compact ion accelerators both with existing high intensity lasers and with future laser systems. Several promising applications are detailed, as well as some of the necessary developments to obtain these results. It is shown that the TNSA mechanism is robust and used for existing applications and that the transparency regime and low density laser ion acceleration could be used for applications requiring high ion energies.

The applicability perspectives offered by these beams are interesting and will be pushed further, in particular thanks to the progresses that can be anticipated for the lasers and the close opening of new installations like ILE, Gemini (at RAL), and further in time Apollon, ELI and diode pumped lasers with very high repetition rate and high energy. Thanks to these installations, it will be possible to increase the maximum energy and the dose of the ions (i) either by extrapolating simply using the mechanisms known nowadays, (ii) or by using other mechanisms (predicted or to be discovered) that will be more efficient.

To do so, the understanding of the accelerating parameters and the optimization of the laser parameters need to improved, by coupling theory and experiments, in order to obtain proton beams of a few hundred MeV, reproducible, controllable and predictable. This will require in particular to explore the scaling laws of existing mechanisms and to test the various mechanisms supposed to take place at ultra-high intensities $10^{21}$-$10^{22}$ W/cm$^2$ or higher (acceleration at the back surface, direct acceleration by the ponderomotive force, acceleration by electrostatic shock, acceleration by radiation pressure).

Future installations will possess various « lines », coupling laser pulses with similar intensities but with very variable durations ultra-short (10 fs) or moderately short (500 fs or more), which will offer the possibility to couple interaction regimes that are different, or even separated, and to take advantage of this coupling. For instance, the adjunction of bright X-ray sources and high energy proton sources will allow to probe dense plasmas heated by the protons. The proton sources will also be used to probe electron acceleration mechanisms. On these installations, it will be important to be able to change the parameters (pulse duration, energy, focalization) because if some applications require an excellent emittance, for others on the other hand the priority will go to ultra high energies (GeV and beyond), or to a maximum number of well collimated ions at low energy, different characteristics that require different ranges of laser parameters.

The progresses mentioned, the increase of energy in particular, should also open the way, as already mentioned, to new applicability perspectives, like injectors for high energy accelerators, the physics of ion beams (injection of multi-charged ion beams in matter), pulsed neutron sources, or even medical applications as well as material spallation and transmutation. Ultra high laser intensities (> $10^{24}$ W/cm$^2$) will even allow to test laser wakefield acceleration concepts, that are nowadays limited to electrons [Tajima and Dawson, 1979], with ions.

An intense international competition is underway both on the modeling of new acceleration regimes and on the experimental validation of these regimes. The recent results obtained at Los Alamos National Laboratory using some of the results presented in this chapter are an important illustration of this.

## Author details

Emmanuel d'Humières

*Université de Bordeaux – CEA - CNRS - CELIA, France*

## Acknowledgement

The author would like to thank Prof. V. T. Tikhonchuk, Prof. Y. Sentoku and Dr. J. Fuchs for fruitful discussions.

## 8. References

Adam J.C. et al. (2006), Phys. Rev. Lett. 97, 205006.
Albertazzi B. et al. (2012), submitted to Eur. Phys. J. Web of Conf.
Albright B.J. et al. (2007), Phys. Plasmas 14, 094502.
Allen M. et al. (2004), Phys. Rev. Lett. 93, 265004.
Alfvèn, H. (1939). On the Motion of Cosmic Rays in Interstellar Space, Phys. Rev. 55, 425.
Andresen G. B. et al. (2007), Phys. Rev. Lett. 98, 023402.
Antici P. et al. (2007), Phys. Plasmas 14, 030701.
Antici P. et al. (2008a), Phys. Rev. Lett. 101, 105004.
Antici P. et al. (2008b), J. Appl. Phys. 104, 124901.
Antici P. et al. (2009), New J. Phys. 11, 023038.
Atzeni S. et al. (2002), Nucl. Fusion 42, L1.
Begay F. et al. (1982), Phys. Fluids 25, 1675.
Bin J. H. et al. (2009), Phys Plasmas, 16, 043109.
Blanchot N. et al. (2008), Plasma Phys. Control. Fusion 50, 124045.
Borghesi M. et al. (2002), Phys. Plasmas 9, 2214.
Borghesi M. et al. (2004), Phys. Rev. Lett. 92, 055003.
Borghesi M. et al. (2006), Fusion Science and Technology 49, 412.
Brunel F. (1987), Phys. Rev. Lett. 59, 52.
Buffechoux S., et al. (2010), Phys. Rev. Lett. 105, 015005.
Bulanov S. V. et al. (2002a), Plasma Physics Reports 28, 453.
Bulanov S. V. et. al. (2002b), Physics Letters A 299, 240.
Bulanov S. V. et al. (2007), Phys. Rev. Lett. 98, 049503.
Bulanov S. V. (2009), Plasma Phys. Control. Fusion 48, B29.
Capdessus R. et al. (2012), submitted to Phys. Rev. E.
Carrié M. et al. (2011), High Energy Density Physics 7, 353.
Ceccotti T. et al. (2007), Phys. Rev. Lett. 99, 185002.
Chen Z. L. et al. (2005), Phys. Rev. E 71, 036403.
Chen H. et al. (2009), Phys. Rev. Lett. 102, 105001.
Chen H. et al. (2010), Phys. Rev. Lett. 105, 015003.
Chen M. et al. (2011), Plasma Phys. Control. Fusion 53, 014004.
Chen S. et al. (2012), Phys. Rev. Lett. 108, 055001.
Clark E. et al. (2000a), Phys. Rev. Lett. 84, 670.
Clark E. et al. (2000b), Phys. Rev. Lett. 85, 1654.
Cowan T. et al. (2002), AIP Conference Proceedings 647, 135.
Cowan T. et al. (2004), Phys. Rev. Lett. 92, 204801.
d'Humières E. et al. (2005), Phys. Plasmas 12, 062704.

d'Humieres E. et al (2006), AIP Conference Proceedings 877, pp. 41-50.
d'Humières E. et al. (2007), Chinese Optics Letters, Vol. 5, Supplement, S136.
d'Humières E. et al. (2010a), J. Phys.: Conf. Ser. 244, 042023.
d'Humières E. et al. (2010b), AIP Conf. Proc. 1299, 704.
d'Humières E. et al. (2011a), MOP153, PAC11 proceedings.
d'Humières E. et al. (2011b), Eur. Conf. Abstracts 35G, P5.005.
d'Humières E. et al. (2012), submitted to Eur. Phys. J. Web of Conf.
Davis S. et al. (2010), Journal of Physics: Conference Series 244, 042006.
Davis S. et al. (2012), submitted to Eur. Phys. J. Web of Conf.
Debayle A. et al. (2010), Phys. Rev. E 82, 036405.
Denavit J. (1992), Phys. Rev. Lett. 69, 3052.
Disdier L. et al. (1999), Phys. Rev. Lett. 82, 1454.
Dong Q. et al. (2003), Phys. Rev. E 68, 026408.
Doumy G. et al. (2004), Phys. Rev. E 69, 026402.
Esirkepov T. Zh. et al. (1999), JETP Lett. 70, 82.
Esirkepov T. et al. (2006), Phys. Rev. Lett. 96, 105001.
Estabrook K. et al. (1978), Phys. Rev. Lett. 40, 42.
Fews A. et al. (1994), Phys. Rev. Lett. 73, 1801.
Flippo K. et al. (2008), Physics of Plasmas 15, 056709.
Flippo K. et al. (2010), J. Phys. Conf. Ser. 244, 022033.
Fourkal E. et al. (2002), Medical Physics 29, 2788.
Fritzler S. et al. (2003), Appl. Phys. Lett. 83, 3039.
Fuchs J. et al. (1999), Phys. Plasmas 6, 2569.
Fuchs J. et al. (2003), Phys. Rev. Lett. 91, 255002.
Fuchs J. et al. (2005), Phys. Rev. Lett. 94, 045004.
Fuchs J. et al. (2006), Nature Physics 2, 48-54.
Fuchs J. et al. (2007a), Phys. Plasmas 14, 053105.
Fuchs J. et al. (2007b), Phys. Rev. Lett. 99, 015002.
Fuchs J. et al. (2009), C. R. Physique 10, 176–187.
Gauthier M. et al. (2012), 39th EPS Conference on Plasma Physics proceedings – Stockholm, Sweden.
Gaillard S. et al. (2011), Phys. Plasmas 18, 056710.
Gitomer S. et al. (1986), Phys. Fluids 29, 2679.
Grech M. et al. (2011), New J. Phys. 13 123003.
Grismayer T. et al. (2006), Phys. Plasmas 13, 032103.
Gurevich A. V. et al. (1966), Sov. Phys. JETP 22, 449.
Habara H. et al. (2004), Phys. Rev. E 70, 046414.
Haberberger D. et al. (2011), Nature Physics 7, 2130.
Hatchett S. et al. (2000), Phys. Plasmas 7, 2076.
Hegelich B. M. et al. (2006), Nature 439, 441.
Henig A. et al. (2009), Phys. Rev. Lett. 103, 245003.
Jung D. (2012), Ph.D. dissertation – Ludwig Maximilians Universität München, Ion acceleration from relativistic laser nano-target interaction.

Kaluza M. et al. (2004), Phys. Rev. Lett. 93, 045003.

Kar S. et al. (2009), Phys. Rev. Lett. 102, 055001.

Kar S. et al. (2011), Phys. Rev. Lett 106, 225003.

Kato T. et al. (2008), Astrophys. J. 681, L93.

Key M. et al. (2006), Fusion Science and Technology 49, 440.

Klimo O. et al. (2008), Phys. Rev. Special Topics-Accelerators And Beams, 11(3):031301.

Klimo O. et al. (2011), New J. Phys. 13, 053028.

Kodama R. *et al.* (2004), Nature 432, 1005.

Krushelnick K. et al. (1999), Phys. Rev. Lett. 83,737.

Krushelnick K. et. al. (2000), IEEE Trans. Plasma Sci 28, 1184.

Kuramitsu Y. et al. (2011), Phys. Rev. Lett. 106, 175002.

Lancaster K. et al. (2007), Phys. Rev. Lett. 98, 125002.

Ledingham K.W. D. et al. (2004), J. Phys. D: Appl. Phys. 37 2341.

Leemans W. et al. (2006) Nat. Phys. 2, 696.

Lefebvre E. et al. (1997), Phys. Rev. E 55, 1011.

Lefebvre E. et al. (2006), J. of App. Phys. 100, 113308.

Limpouch J. et al. (2008), Laser & Part. Beams 26, 225.

Macchi A. et al. (2009), Phys. Rev. Lett. 103, 085003.

Mackinnon A.J. *et al. (2002)*, Phys. Rev. Lett. 88, 215006.

Mackinnon A.J. et al. (2004), Rev. Sci. Inst. 75, 3531.

Malka G. *et al. (1996)*, Phys. Rev. Lett. 77, 75.

Malka V. et. al. (2004), Medical Physics 31, 1587.

Malka V. (2008), Principles and applications of compact laser–plasma accelerators, Nature
    Physics 4, 447-453.

Maksimchuk A. et al. (2000), Phys. Rev. Lett. 84, 4108.

Mancic A. *et al. (2010a)*, High Energy Density Physics 6, 21.

Mancic A. et al. (2010b), Phys. Rev. Lett. 104, 035002.

Mora P. (2003), Phys. Rev. Lett. 90, 185002.

Mora P. (2005), Phys. Rev. E 72, 056401.

Mourou G. et al. (2006), Rev. Modern Phys. 78, 309.

Murakami Y. et al. (2001), Phys. Plasmas 8, 4138.

Nakamura T. et al. (2004), Phys. Rev. Lett. 93, 265002.

Nakamura T. et al. (2007), Phys. Plasma 14, 103105.

Nakatsutsumi M. et al. (2007), Phys. Plasma 14, 050701.

Nakatsutsumi M. et al. (2008), J. of Physics: Conf. Series 112, 022063.

Naumova N. et al. (2009), Phys. Rev. Lett. 102, 025002.

Neely D. et al. (2006), App. Phys. Lett 89, 021502.

Nemoto K. et al. (2001), Appl. Phys. Lett. 78, 595.

Offermann D.T. et al. (2011), Phys. Plasmas 18, 056713.

Palmer C.A. et al. (2011), Phys. Rev. Lett. 106, 014801.

Patel P. et al. (2003), Phys. Rev. Lett. 91, 125004.

Peano F. et al. (2007), Phys. Plasmas 14, 056704.

Pelka A. et al. (2010), Phys. Rev. Lett. 105, 265701.

Perez F.et al. (2010a), Phys. Rev. Lett. 104, 085001.
Perez F. et al. (2010b), Phys. Plasmas 17, 113106.
Pommier L. et al. (2003), Laser Part. Beams 21, 573.
Popescu H. *et al. (2005)*, Phys. Plasmas 12, 063106.
Psikal J. et al. (2008), Phys. Plasmas 15, 053102.
Psikal J. et al. (2010), Phys. Plasmas 17, 013102.
Pukhov A. (2001), Phys. Rev. Lett. 86, 3562.
Quéré F. et al. (2006), Phys. Rev. Lett. 96, 125004.
Renard-Le Galloudec N. et al. (2008), Rev. Sci. Inst. 79, 083506.
Renard-Le Galloudec N. et al. (2009), Phys. Rev. Lett 102, 205003.
Renard-Le Galloudec N. et al. (2010), Laser and Particle Beams 28, 513.
Robinson A. P. L. et al. (2007), Phys. Plasmas 14, 083105.
Robinson A. P. L. et al. (2008), New Journal Of Physics, 10:013021.
Robson L. et al. (2007), Nat. Phys. 3, 58.
Romagnani L. *et al.* (2005), Phys. Rev. Lett. 95, 195001.
Roth M. et al. (2001), Phys. Rev. Lett. 86, 436.
Roth M. *et al.* (2002), Phys Rev ST-AB 5, 061002.
Ruhl H. et al. (2001), Plasma Phys. Report, 27, 5, 363-371.
Ruhl H. et al. (2004), Phys. Plasmas 11, L17.
Sack C. et al. (1987), Physics Reports 156, 311.
Santala M. et al. (2000), Phys. Rev. Lett. 84, 1459.
Santala M. et al. (2001), Appl. Phys. Lett. 78, 19.
Sarkisov G. S. et al. (1999), Phys. Rev. E 59, 7042.
Schirber M. et al. (2005), Science 310, 1610-1611.
Schlegel T. et al (2009)., Phys. Plasmas 16, 081303.
Schollmeier M. et al. (2011), Bull. Am. Phys. Soc.
    http://meetings.aps.org/link/BAPS.2011.DPP.NO7.1.
Schwoerer H. et al. (2006), Nature 439, 445.
Sentoku Y. (2000) et al., Phys. Rev. E 62, 7271.
Sentoku Y. et al. (2003a), Phys. Plasmas 10, 2009.
Sentoku Y. et al. (2003b), Phys. Rev. Lett. 90, 155001.
Sentoku Y. et al. (2004), Phys. Plasmas, 11, 3083.
Sentoku Y. et al. (2008), J. Comput. Phys. 227, 6846.
Sentoku Y. et al. (2011), Phys. Rev. Lett. 107, 135005.
Silva L. et al. (2004), *Phys. Rev. Lett.* 92, 015002.
Snavely R. et al. (2000), Phys. Rev. Lett. 85, 2945.
Snavely R. et al. (2007), Phys. Plasmas, 14, 092703.
Sokolov I. V. (2009a), Journ. Exp. Theor. Phys. 109, 207.
Sokolov I.V. et al. (2009b), Phys. Plasmas 16, 093115.
Sokolov I.V. et al. (2010), Phys. Rev. E 81, 036412.
Sokolov I. V. et al. (2011), Phys. Plasmas 18, 093109.
Solodov A.A. et al. (2009), Phys. Plasmas 16, 056309.
Solodov A. A. et al. (2010), J. Phys. Conf. Series 244, 022063.

Spitkovsky A. (2008), Astrophys. J. 673, L39.

Stephens R. et al. (2004), Phys. Rev. E 69, 066414.

Storm M. et al. (2009), Phys. Rev. Lett. 102, 235004.

Tabak M. et al. (1994), Phys. Plasmas 1, 1626.

Tabak M. et al. (2006), Fusion Sci. and Tech. 49, 254.

Tajima T. and Dawson J. (1979), Phys. Rev. Lett. 43, 267–270.

Tamburini, M. et al. (2010), New Journ. Physics 12 123005.

Tamburini M. et al. (2012), Phys. Rev. E 85, 016407.

Temporal M. et al. (2002), Phys. Plasmas 9, 3098.

Temporal M. (2006), Phys. Plasmas 13, 122704.

Tikhonchuk V.T. et al. (2005), Plasma Phys. and Control. Fusion 47, 869.

Toncian T. et al. (2006), Science 312, 410.

Weibel E.S. (1959), Phys. Rev. Lett. 2, 83.

Wilks S. et al. (1992), Phys. Rev. Lett. 69, 1383.

Wilks S.C. (1993), *Simulations of ultraintense laser-plasma interactions*, Phys Fluids B 5, 2603.

Wilks S. C. et al. (2001), Phys. Plasmas 8, 542.

Willingale L. et al. (2006), Phys. Rev. Lett. 96, 245002.

Willingale L. et al. (2007), Phys. Rev. Lett. 98, 049504.

Willingale L. et al. (2009), Phys. Rev. Lett. 102, 125002.

Wittmann T. et al. (2006), Rev. Sci. Instrum. 77, 083109.

Yamagiwa M. et al. (1999), Phys. Rev. E 60, 5987.

Yang J. M. et al. (2004), J. Appl. Phys. 96, 6912.

Yin L. et al. (2006), Laser and Particle Beams 24, 291.

Yogo A. et al. (2008), Phys. Rev. E 77, 016401.

Youssef A. et al. (2006), Phys. Plasmas 13, 030702.

# Chapter 3

# GeO₂ Films with Ge-Nanoclusters in Layered Compositions: Structural Modifications with Laser Pulses

Evgenii Gorokhov, Kseniya Astankova, Alexander Komonov
and Arseniy Kuznetsov

Additional information is available at the end of the chapter

## 1. Introduction

In this chapter, we will discuss issues related to the development of new materials and device technology for micro-, nanoelectronics and optics. More specifically, our research activities were devoted to the study of objects in which 3D quantum size effect was revealed. These are a group of materials consisting of indirect-band-gap semiconductor nanoclusters embedded in insulator (Knoss, 2008; Molinari et al., 2003; Takeoka et al., 1998).

Changes in the optical properties of Ge-nanoclusters due to 3D quantum size effect in metastable germanium monooxide (GeO(solid)) layers after decomposition of such layers into Ge and $GeO_2$ ($GeO_2$<Ge-NCs>) were observed by us as early as in the late seventies, i.e. a few years before this effect was for the first time reported in the literature (Ekimov & Onuschenko, 1981). It should be emphasised here that in our experiments this effect was observed in a thin-film heterosystem rather than in bulk $SiO_2$ glasses with CuCl or CdS precipitates. The latter circumstance is important for using such materials in modern film technology of micro-, nano- and optoelectronics.

Apart from the detection of the photoluminescence coming from $GeO_2$<Ge-NCs> heterolayers, we showed that the 3D quantum size effect, radically changing the properties of the electron subsystem of the solid, could also be used to achieve a dramatic modification of the lattice subsystem of the solid matter. So, a new material, nanofoam, a solid similar to aerogels (Hrubesh & Poco, 1995), was obtained from germanium dioxide in $GeO_2$<Ge-NCs> heterolayers (Gorokhov et al., 2011). Simultaneously, we have developed a new technique allowing easy production of Ge-quantum dots (Ge-QDs) of a very small size. The dispersion of QDs sizes in obtained $GeO_2$<Ge-NCs> layers could be significantly reduced in comparison with the dispersion of Ge-QD sizes in $GeO_2$<Ge-NCs> heterolayers.

In our work, thin films from the following germanium oxides were investigated:

1.  GeO(solid) films, which were layers of (usually amorphous) stoichiometric germanium monooxide (also, there are reported data about the existence of crystalline modifications of GeO(solid) (Martynenko et al., 1973));
2.  $GeO_x$ films, which were nonstoichiometric GeO(solid) layers with x, standing to indicate the chemical composition of the material, ranging in $0 < x < 2$ (I) or $1 < x < 2$ (II). In case I, the composition of $GeO_x$ varies from pure germanium to $GeO_2$, and in case II, it varies between the stoichiometric compositions of GeO(solid) and $GeO_2$;
3.  $GeO_2$ films, which can be amorphous or have crystalline modifications;
4.  heterogeneous material $GeO_2<Ge-NCs>$, obtained from metastable GeO(solid) during its chemical decomposition into two components: an amorphous $GeO_2$ matrix and Ge-nanoclusters dispersed throughout the matrix.

These film materials, belonging to dielectrics, were recognised inappropriate for use in planar Si-based IC technology, and they were therefore forgotten. The main reason was that germanium monooxide was a metastable material readily undergoing decomposition even without any additional heating (Marin, 2010). Second, the layers of amorphous germanium dioxide rapidly dissolve in water (Kamata, 2008). Therefore, layers of amorphous GeO(solid) and $GeO_2$ were recognised inferrior to thermal silicon dioxide films, as well as to high-temperature silicon nitride films, which also proved to be chemically resistant and mechanically strong. In addition, it was absolutely unclear how patterning or selective etching of germanium oxide layers could be achieved. Other layers of germanium oxides, such as hexagonal $GeO_2$ and $GeO_2<Ge-NCs>$ heterolayers, were unknown at that time when the possibility of using germanium oxide films in semiconductor industry was under evaluation. Also, there were no reported data on the interaction of layers of amorphous $GeO_2$ with $SiO_2$ or $Si_3N_4$ and on subsequent crystallisation of such binary compositions (Gorokhov et al., 1987, 1998).

However, the study of properties and specific features of all modifications of film compounds based on germanium oxides have led us to a revision of the appropriateness of such compounds for use in modern technology of solid-state devices in micro-, nano- and optoelectronics. The main advantage of germanium oxide layers over other dielectric layers consists in the capability of such layers (both atomic and electronic subsystem) to be easily modified during treatments given to the layers, and the modifications lead to a significant change of the initial properties of the material. Therefore, a deposited film of germanium oxides should now not be considered as a fully complete and final result of some process sequence; instead, it should be considered as a material suitable for subsequent modifications. In particular, based on such layers, one can create light-emitting diodes, photodiodes, optoelectronic couples, optical fibers, interference filters, mirrors, lenses, diffraction gratings and holograms, as well as single electron transistors, memory elements, resists for laser and probe nanolithography, low-k and high-k dielectrics, and a component of colloidal solutions of Ge-nanoparticles to fight against cancer in medicine (Tyurnina et al., 2011). To date, not all device applications of germanium oxide layers have been identified.

From our point of view, the specific properties of germanium oxide compounds outlined above show great promise in the development of the nanotechnology of layered systems in nanoelectronics. A pressing problem in this field is developing methods for local modification of film coatings aimed at imparting desired properties to a small area of the film only. Especially, there is a problem of formation of elements having extremely small sizes in one or two dimensions in a thin continuous planar layer.

Studying the potential offered by narrow linear and dot scanning laser treatments as applied to thin-film materials or surfaces exhibiting photosensitivity or susceptible of high-intensity focused laser radiation, i.e. laser micro- and nanolithography, was one of the promising research lines at this topic. A big volume of research has been done by us in this field, and that has brought forth many interesting effects and successful developments. Achievement of modification effects in thin near-surface layers of substance while leaving the material intact at larger depths is another research area in using laser treatments in micro- and nanoelectronics. Local modifications of material properties could also be achieved in the volume of bulk samples using a laser beam focused at a point inside the sample, and then the material remains non-modified on the surface and in subsurface layers.

Obviously, for a stronger display of necessary effects, special materials should be used in such treatments. Easy initiation of processes leading to structural modification of materials is the main property of required layers. The lower is the energy threshold of such modifications in a substance, the easier is the initiation of the process, and the stronger is this effect induced by the treatment. There exist substances in which processes leading to a change of their structure and chemical composition can be activated with an increase of the level of specific energy introduced in unit volume of the substance. Each of such processes is characterised by its own activation energy, and the activation energies of different processes normally differ in value quite widely so that the processes proceed separately. Of course, the more such potential processes can proceed in a material and the lower are the activation energies of the processes, the wider is the diversity of forms and transformations that can be induced in such an initial material.

Metastable germanium monoxide layers can serve such a material. The capability of such layers to transformation into a chemically and structurally stable germanium dioxide allows us to include modifications typical of GeO$_2$ in a number of possible modifications of initial GeO(solid) layers. During the decomposition of GeO(solid) layers, atomic germanium forms quantum-sized Ge-nanoparticles; this process also adds to the potential of possible transformations.

The two unique properties of film systems based on germanium oxides, their capability to easy modification of their electron and lattice subsystems, are well complemented with a third unique capability – easy transformation of material properties under pulsed laser irradiation. Local laser pulse treatments of the samples were found to be a technique enabling easy modification of germanium oxide layers. This technique offers us a unique tool for realising the potential inherent to germanium oxide layers. Thus, the content of this chapter aims to acquaint the reader with the main results of our investigations in the indicated field.

## 2. Experimental methods

### 2.1. Methods of synthesis of GeO(solid) layers and GeO₂<Ge-NCs> heterolayers

The studied films were obtained using three film deposition methods. First, heterolayers of GeO₂<Ge-NCs> were deposited from supersaturated GeO vapour in a low-pressure chemical vapour deposition (LP CVD) process (Knoss, 2008) onto substrates located in a quartz flow reactor (Fig. 1). The implemented LP CVD process involves two stages. The first stage (Fig. 1, zone A) is the formation of germanium monoxide molecules (GeO(gas)) according to the reactions:

$$2Ge(solid) + O_2(gas) \rightarrow 2GeO(gas) \tag{1}$$

$$Ge(solid) + H_2O(gas) \rightarrow GeO(gas) + H_2(gas) \tag{2}$$

The second stage (Fig. 1, zone B) involves two subsequent processes (a and b) proceeding during deposition of supersaturated GeO(gas) onto substrates. Process a is the condensation of vapour molecules (GeO(gas) → GeO(solid)) proceeding with the formation of a homogeneous metastable solid layer of germanium monoxide (GeO(solid)) on the substrate. The solid GeO film is metastable, and it readily decomposes into Ge and GeO₂ (process b) in several minutes at relatively low temperatures about 300°C and over:

$$GeO(solid) \rightarrow \tfrac{1}{2}Ge + \tfrac{1}{2}GeO_2 \tag{3}$$

In reaction (3), the germanium dioxide GeO₂ forms a glassy matrix, with excess germanium atoms being segregated as Ge-nanoparticles. The film growth rate depends on GeO vapour pressure and substrate temperature. In this way, using reactions (1) – (3) we were able to obtain either GeO₂ films with Ge-nanocrystals with sizes ranging from ~ 2 nm to ~ 10 nm (higher deposition temperatures, see Fig. 1 b, area III), or GeO₂ films with amorphous Ge-nanoclusters (lower deposition temperatures, Fig. 1 b, area II), or GeO(solid) films (deposition at room temperature, Fig. 1 b, area I). It should be emphasised here that the heterostructures used in our experiments had the following remarkable property: their molar ratio between Ge and GeO₂ was always fixed at exactly 1:1 independently on particular implemented growth conditions. The surface density of Ge-nanoclusters in a single-layer coating could range from ~$10^{10}$ to ~$10^{14}$ NC/cm², the average distance between Ge-nanoclusters being 1/2 of their diameter.

To obtain thin, stoichiometric, nondecomposed and homogenous GeO films on various substrates, a second method was employed. GeO(solid) films were additionally deposited onto substrates using thermal re-evaporation in a high-vacuum ($10^{-7}$ Pa) flow reactor of thick GeO₂<Ge-NC> heterolayers (400–500 nm) grown by the first method (Sheglov, 2008). In such a process, a sample with a GeO₂<Ge-NCs> heterolayer was heated to a temperature of 550–600 °C by an ohmic heater. The heated GeO₂<Ge-NCs> film evaporated in accordance with the reverse of the deposition reaction:

$$\tfrac{1}{2}GeO_2(solid) + \tfrac{1}{2}Ge(solid) \rightarrow GeO(gas)$$
(4)

The resulting GeO vapour condensed onto a cold substrate ($T \sim 25$ °C), yielding a layer of metastable solid germanium monoxide GeO(solid): GeO(gas) → GeO(solid) (Fig.2). At temperatures $T > 250$ °C the GeO(solid) films decomposed, according to reaction (3), into Ge and GeO₂ with the formation of a new GeO₂<Ge-NCs> heterolayer.

In the third method (Ardyanian et al., 2006), using evaporation of GeO₂ with an electron beam in a high-vacuum chamber followed by deposition of evaporated species onto substrates at a low temperature (100 °C), we were able to obtain non-decomposed GeOₓ films with $x = 1,2$. Those films, of thicknesses about 300 nm, were obtained in Laboratoire de Physique des Matériaux (LPM), Nancy-Université, CNRS, France. In both cases, with the help of the decomposition of GeO(solid) proceeding according to reaction (3), we could obtain layers of GeO₂ with embedded Ge-nanoclusters.

All studied films were deposited onto either Si (100) substrates or glass. To avoid evaporation of the GeO(solid) films during laser treatments, the samples were covered with protective SiO₂ or SiNₓOy layers. Protective SiNₓOy cap layers about 25 nm thick were deposited at 100 °C using the plasma enhanced chemical vapour deposition method (PE CVD), in Institute of Semiconductor Physics of the Siberian Brunch of Russian Academy of Science (ISP SB RAS). Protective SiO₂ cap layers were prepared using two methods. The first method was the evaporation of fused silica glass with an electron beam at 100 °C in a high vacuum ($10^{-8}$ Torr); this method was implemented at Nancy-Université, France (Jambois et al., 2006). The second method was chemical vapour deposition of SiO₂ layers with simultaneous monosilane oxidation in the pressure range 0,5-1,2 Torr at 150°C; this method was developed and implemented at ISP SB RAS.

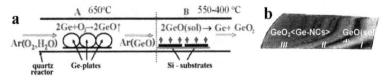

**Figure 1.** The scheme of LP CVD process of GeO₂<Ge-NCs> heterolayer growth (a); and photo of the GeO₂<Ge-NCs> film on Si-substrate of gradient thickness, structure and features (b).

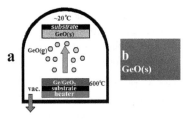

**Figure 2.** Scheme of vacuum GeO(solid) formation (a); and photo of the GeO(solid) film of uniform properties on Si-substrate (b).

## 2.2. Methods of structural analysis

Investigation into the structural properties of thin layers of amorphous dielectrics, and also investigation into the impact of various growth factors and technological treatments on those properties, present most challenging problems in modern material science for micro- and nanoelectronics. This usually requires a complex of specific instruments and methods. We examined our films using Raman scattering spectroscopy, IR spectroscopy, ellipsometry, scanning electron microscopy (SEM), transmittance electron microscopy (TEM), and atomic force microscopy (AFM).

The optical constants of our films were studied with the help of the scanning ellipsometry method, implemented with step $\delta l$=0,5 mm, using a He-Ne laser (633 nm). The results were interpreted using a specially developed algorithm described elsewhere (Marin et al., 2009). The spectral dependences of absorption and refraction indexes, $k(D)$ and $n(D)$, versus the diameter of Ge-nanoclusters (D) contained in the films were studied using many-angle, multiple-thickness spectral (250-800 nm) scanning ellipsometry. With the combination of spectral many-angle, multiple-thickness measurements, we were able to significantly improve the accuracy of ellipsometric data on the thickness and optical constants of studied films in the visible range.

Atomic force microscopy (Solver P-47H, NT-MDT, Russia) and scanning electron microscopy (Hitachi S4800) were used to examine the surface morphology of our films before and after applied modifications. Optical microscopy was used to register changes in the optical constants of the films in laser-modified areas. The direct observation of Ge-nanoclusters in $GeO_2$ films was carried out with the help of transmission electron microscopy (TEM) on specially prepared thin $SiO_2$ membranes (Gorokhov et al., 2006).

Raman spectroscopy was used to reveal and identify the structure (amorphous or crystalline) of Ge-nanoclusters in the films. Raman spectra were registered in quasi back-scattering geometry with a 514,5-nm $Ar^+$ laser used as an excitation source. A double DFS-52 spectrometer and a triple T64000 Horiba Jobin Yvon spectrometer equipped with a micro-Raman setup were employed. In the latter case, slightly unfocused laser beam was used, producing a spot of about 3-4 micrometer diameter on the surface of the sample, with the laser power reaching the sample being diminished to 10-20 mW in order to avoid overheating of the films. All Raman spectra were measured at room temperature.

IR spectroscopy (FT-801 Fourier spectrometer, ISP SB RAS, Russia, equipped with a micro-setup) was used to study the chemical composition and structure of dielectric layers after different treatments given to them in small areas of the sample. IR absorption measurements carried out at normal incidence were performed at a resolution of 4 cm$^{-1}$.

Treatments of GeO(solid) layers were performed using a nanosecond KrF excimer laser (wavelength 248 nm, pulse duration 25 ns). Laser fluences ranging from 130 to 170 mJ/cm$^2$ were applied. A Ti-Sapphire laser (FemtoPower Compact Pro, Femtolasers Produktions GmbH) with 800 nm central wavelength and pulse duration < 30 fs was used for laser treatments. The energy distribution in the laser spot was assumed to have a Gaussian form.

The laser fluence was changed by varying the laser pulse energy $E_{pulse}$: $E_0 = 2E_{pulse}/\pi r_0^2$, where $E_0$ is the maximum energy and $r_0$ is the laser spot radius. The treatments were done in scanning mode with the laser spot diameter being 70 or 1 μm.

## 3. Peculiarities of GeO(solid) films and GeO$_2$<Ge-NCs> heterolayers

### 3.1. The structure and the decomposition process of the metastable GeO(solid) layers

Because of the weak interest of researchers to the films of germanium oxides, nowadays in the scientific literature there are no unambiguous data about the structure of GeO(solid) layers, about the mechanisms of formation of such layers under different conditions, and about their decomposition according to the reaction (3) or evaporation of decomposed GeO(solid) films under heating according to the reaction (4) (Kamata, 2008). However, for controllable variations of the GeO(solid) layers structure and properties, adequate to reality considerations on these aspects are required. This urged us to look for answers to these questions. The qualitative considerations described below are different in that its main statements do not contradict to any fact from known experimental data on GeO(solid) layers. By analogy, one can extend this consideration to the description of the behaviour and properties of SiO(solid) films, since Ge and Si are both group IV elements, this circumstance makes compounds of germanium and silicon with other elements quite similar.

The technology of the films consisting of Si- or Ge-QDs introduced in a dielectric SiO$_2$ or GeO$_2$ matrix, respectively, demands understanding of the role and properties of SiO and GeO monoxides. They can be solid and gaseous. However, the physical chemistry of Ge and Si lower oxides is studied insufficient. The actuality of their studies originates from the fact that GeO(gas) condensation in the form of a thin GeO(solid) film is the simplest way of obtaining Ge-QDs dispersed in the glassy GeO$_2$ film. As the GeO(solid) structure is thermodynamically unstable, the film quickly decomposes due to reaction GeO(solid) → ½Ge + ½GeO$_2$. To obtain vapour from GeO molecules, one can easy use inverse reaction ½Ge + ½GeO$_2$ → GeO(gas) at T > 500 °C (Knoss, 2008). The produced heterolayers with a high Ge-QDs concentration in GeO$_2$ matrix have the peculiarities which are interesting from both scientific and practical standpoints (Knoss, 2008). For controllable variations of the GeO$_2$<Ge-NCs> heterolayers structure (in particular, of Ge-nanoparticles characteristics), reliable control of the decomposition process of GeO(solid) layers and a good understanding of the nature of its thermodynamic metastability are required. So we had to look for an explanation to one of the unsolved properties of solid germanium monoxide – thermodynamic metastability of GeO(solid).

In studying germanium monoxide, there arises a question why GeO molecules in gaseous state are stable (dissociation energy $\Delta H^o$ ~159 kcal/mole (Jolly & Latimer, 1952)), whereas in solid GeO germanium atoms turn out to be capable of easily breaking their bonds with oxygen atoms without any activation at room temperature. The difference in the stability of the two, gaseous and solid, aggregate states of GeO can be attributed to the fact that the Ge- and O-atoms in a GeO vapour molecule have no immediate interaction with other atoms.

Obviously, sp³-hybridization (tetrahedral orientation of the four orbitals of Ge-atom) in this molecule is not possible. Therefore in a GeO(gas) molecule the Ge-atom is bonded with the O-atom with one σ-bond and one π-bond (Tananaev & Shpirt et. all, 1967). Transformation of Ge-atom orbitals from sp³-hybridized state (inherent to solid Ge and GeO₂) into the structure of valence bonds inherent to GeO(gas) molecules requires energy $\Delta H_{298}\sim54$kcal/mole (Jolly & Latimer, 1952) and heating.

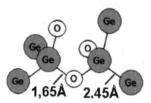

**Figure 3.** Assumed structure of the atomic network of GeO(solid) satisfying the following two conditions: the valence orbitals of Ge-atoms are *sp³*-hybridized and the chemical composition of the GeO(solid) film complies with the Ge-O stoichiometry.

When diatomic GeO(gas) molecules condense, the distances between individual GeO molecules become the order of the atomic size. As a result, Ge- and O-atoms become able to be bonded into 3D network from chains of Ge- and O-atoms. The latter allows sp³-hybridization with tetrahedral orientation of the valence orbitals of Ge-atoms to form the base for construction of an atomic network. The driving force for such a transformation can be the release of free energy spent on the formation of a new valence bond structure of GeO(gas) molecules (reaction (4)). Thus, as the lattices of germanium dioxide and germanium, the lattice GeO(solid) should consist of linked tetrahedra. According to the stoichiometry of GeO each Ge-atom at the vertex of a tetrahedron, formed by its sp³-hybridized bonds, should be bonded to two other Ge-atoms and two O-atoms (see Fig. 3).

However, there are differences between the lengths of Ge-Ge (2,45 Å) and Ge-O (1,73 Å) bonds in the elemental tetrahedrons of the GeO(solid) lattice. Therefore, this lattice is heavily deformed and has strongly distorted bond angles. So, the real structure of GeO(solid) is not compact, its lattice energy is not minimised to ensure the stability of the atomic network. The latter situation is typical of materials and their basic lattice elements have a low degree of symmetry. When such elements become arranged in a continuous atomic network by successive translations, large cavities should appear in the film. To make such cavities filled with the material, strong deformations of the translation elements, tetrahedrons, become necessary. In any case, the formed atomic network possess a high level of internal strain energy. This will force GeO(solid) to decompose into components, the structure of such components is constructed from (a) basic elements with a higher degree of symmetry and based on sp³-hybridization of Ge-atom orbitals (b). I.e. these basic elements must provide a layering of unstable material Ge(solid) into the components made without the high internal strain energy. Such components are Ge and GeO₂ which have stable lattices based on elements with high degree of symmetry – Ge(Ge³⁻)₄ and Ge(O⁴⁺)₄ tetrahedra, respectively.

A model for the atomic structure of GeO(solid) were developed on data received by IR and Raman spectroscopy, scanning probe microscopy and new methods in high-resolution TEM and ellipsometry. The model allows us to consciously control the structure of heterolayers GeO₂<Ge-NCs> during their growth and further treatment. An analysis of structural transformations of germanium monoxide based on this model allows gaining a better insight into some important details of the formation of the metastable GeO(solid) atomic network during condensation of GeO(gas) molecules and of decomposition under activating treatments. According to the above model, the formation of stable GeO₂<Ge-NCs> heterolayers proceeds in two steps. The first step is the formation of a homogeneous metastable GeO(solid) structure at T< 100-200°C, when GeO(gas) molecules condense from vapour onto the surface of the growing GeO(solid) film. It is a very rapid process (femtoseconds) proceeding when the atomic bonds of Ge- and O-atoms in absorbed GeO(gas) molecules get rearranged into tetrahedral configuration of sp³-hybridized bonds; the process has a very high rate and the duration of this process is defined by the switching time of valence bonds between chemically interacting atoms.

Note that if during the condensation of GeO(gas) molecules its diatomic molecules cannot approach each other at atomic distances, the structure of the valence bonds of those molecules remains unaltered. It follows from the data by Ogden and Ricks (Ogden & Ricks, 1970), who found that during condensation GeO(gas) molecules readily become incorporated into the solid matrix of frozen N₂ or Ar at T ~ 20 K. Under such conditions, the vibrational spectra of GeO(gas) molecules (Fig. 4 a) show no changes during freezing in the solid state, and their IR absorption spectra are no difference from the spectra of GeO vapour molecules. In contrast, in the case of condensation of GeO(gas) molecules proceeding with the formation of a GeO(solid) film, vibrational characteristics of the atomic network change considerably (see Fig. 4 b). IR spectrum of frozen GeO molecules has no band, which is typical for GeO(solid) during its thermal decomposition in the range of 770 - 870 cm⁻¹.

The second step is a process of rearrangement of the homogeneous but metastable lattice of GeO(solid) layer into a complicated heterogeneous film structure such as a solid glassy GeO₂ matrix with incorporated Ge-nanoparticles. The latter process is more difficult and by order of magnitude slower than the former process. At this step, two stages can be marked out. The first stage involves decomposition of the lattice of the GeO(solid) layer, with half of the Ge-atoms of this lattice becoming bonded to the bridge O-atoms with all the four sp³-hybridized bonds of each Ge-atom. Here, a solid GeO₂ network made by Ge(O²⁻)₄-type tetrahedrons forms. This network fills up the volume that was previously fully occupied by the 3D GeO(solid) network. On the contrary, the second half of the Ge-atoms in the decaying GeO(solid) network tends to fully disrupt its sp³-hybridized valence bonds with O atoms to get out of the atomic network of solid germanium oxide. However, because of the impossibility to do that immediately, these atoms are forced to form various defects in glassy GeO₂. The variety of atomic configurations in GeO₂ layers could be rather large. But the numerous atomic configurations can be subdivided into two fundamentally different types: type 1 is Ge-atoms at interstitial sites not bonded to the lattice; and type 2 is Ge-atoms partially preserving their bonds with the lattice. In the latter case, such atoms are elements of sp³-hybridized distorted tetrahedra involved in continuous chains formed by undistorted, successively aligned tetrahedra. The distorted tetrahedra are tetrahedra lacking some bridge O-atoms on their tops; i.e. the central Ge-atom of such a tetrahedron has one to three dangling sp³-hybridized valence bonds or those bonds link the central Ge-atom to other Ge-atoms (here, different combinations of the two variants are also possible). Note that during the formation of the $GeO_2$ network from the lattice of GeO(solid), there is no need in a transfer of large masses of substance between remote sites in the volume of the material. Structural rearrangements of the 3D matrix proceed simultaneously in a huge number of centres, and the rearrangement processes proceed in a local vicinity of each centre.

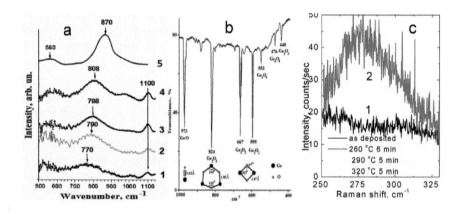

**Figure 4.** a - IR transmission spectra of gaseous GeO molecules (monomer, dimer, trimer, tetramer), embedded in matrix of frozen $N_2$ or Ar at T = 20 K (Ogden & Ricks, 1970); b - IR absorption spectrum of GeO(solid), prepared by condensation of GeO vapour: **1** - as-deposited; **2, 3, 4, 5** - annealed at 260°C/6 min; 290°C/5 min; 320°C/5 min; 600°C/10 min, respectively; c - Raman spectra of GeO(solid): **1**- as deposited; **2** – after series of annealings at 260°C/6 min; 290°C/5 min; 320°C/5 min.

That is why migrations of many atoms turn out to be quite localised processes. However, simultaneous decay of one lattice and formation of the other lattice lead to the formation of a multitude of various defects in the forming atomic network. The density of such defects is several orders higher than the values typical of the equilibrium lattice states. As a result, simultaneously with the intensive formation of defects in the $GeO_2$ network, there may be no less intensive annihilation of such defects and complementary defects. Hence, during structural rearrangement of a GeO(solid) film into a heterostructure, density and viscosity of the material should considerably decrease. Of course, the characteristic times of such rearrangements must be large and depend on the temperature and other external factors. The structural modification process of germanium oxide lattice in the initial GeO(solid) layer subjected to successive anneals is shown in IR spectroscopy data (Fig. 4 b). Film of 55 nm thickness was deposited at T~ 25 °C. After two annealings (260 and 290 °C) shrinkage of layer thickness occurred by ~ 6-8% according to ellisometric data, as a result the structure of discussed heterosystem seeks to optimise its stable parameters: the size, the number of defects in the atomic net, the excess of internal energy, etc.

The above specific features of the rearranging lattice of the oxide layer facilitate the nucleation of Ge-nanoclusters in the volume of the GeO$_2$ matrix, the second component of the structural arrangement process of GeO$_2$<Ge-NCs> heterolayer. Very likely, this process is the slowest one among the above-discussed processes, as Ge-atoms need to lose their continuous bonds with the oxide lattice and, then, migrate over sufficiently large distances (much larger than the size of the "unit cell" in the oxide network) to enter the structure of growing Ge-clusters. The segregation kinetics of amorphous germanium in GeO(solid) layers annealed at different temperatures is reflected in the Raman data shown in Fig. 4 c. Combined Raman spectroscopy and HRTEM data show that germanium initially forms small amorphous Ge-particles sized several nanometers; on increasing the temperature and the duration of the film synthesis process and subsequent anneals, those particles grow in size to form larger Ge-particles sized several ten nanometers. According to Raman data, at T ~ 470-490 $^0$C, amorphous Ge-particles undergo crystallisation. At T > 600$^0$C, in a GeO$_2$<Ge-NCs> heterostructure the GeO$_2$ matrix may also transform into hexagonal phase, which is isomorphous to $\alpha$-quartz (Gorokhov, 2005).

## 3.2. Elementary GeO$_2$ lattice defects in modifications of GeO$_2$<Ge-NCs> heterolayers

Analysis of rearrangement processes proceeding during decay of metastable germanium monoxide layers should be performed considering the fact that such processes are controlled by regularities typical of glasses rather than crystals, as microscopic mechanisms underlying lattice transformation and relaxation processes in glasses and crystals are radically different. For gaining a better insight into substance modification processes proceeding in GeO$_2$<Ge-NCs> heterosystems under pulsed laser treatments, we have to first consider the role of elementary defects in glass atomic network during its formation and subsequent treatments.

In glasses transitions during melting and freezing do not have a sharp boundary. Such processes proceed gradually as the glass viscosity ($\eta$) continuously decreases (upon heating) or increases (upon cooling). Vitrification temperature ($T_g$), at which glass viscosity $\eta = 10^{13}$ P, is accepted as the phase transition point. At T < $T_g$ glassy materials are solid, with the majority of atomic bonds in their bulk being not broken; at T >> $T_g$ the materials are melted, with their atomic bonds undergoing rupture, and that provides the material's easy flow ability. Silicon and germanium oxides are basic natural glass-forming materials, but the vitrification temperatures of pure SiO$_2$ and GeO$_2$ are strongly different ($T_g$(GeO$_2$) =570$^0$C, $T_g$(SiO$_2$) =1170$^0$C), causing a considerable difference in material properties of silicate and germanate glasses. However, the physicochemical mechanisms underlying the behavior and properties of silicate and germanate glasses and the mechanisms controlling structural and chemical modifications of the two groups of glasses, do not have vital differences, although quantitative characteristics defining material properties and processes in glasses are usually considerably different.

The viscous glass flow is a thermally activated process: $\eta(T)=A\cdot exp(Q/RT)$, where $Q$ is the viscosity activation energy and A is a constant. In amorphous materials, the viscous flow clearly deviates from the Arrhenius law: in such materials, the viscosity activation energy $Q$ changes its magnitude from a higher value $Q_H$ at low temperatures (glass state) down to a smaller value $Q_L$ at high temperatures (liquid state). In the theory of glass viscosity ($SiO_2$, $GeO_2$ and others), much attention is normally paid to defect migration in the glass atomic network. Presently available models of this phenomenon are based on the Mueller hypothesis (Mueller, 1955, 1960) about switch-over of oxygen bridge bonds in material. According to this hypothesis, during viscous melt flow, covalent Si-O (or Ge-O in $GeO_2$ glass) bonds do not break (Appen, 1974), but they switch over (translate, reorient). That is why the mechanism of viscous flow in mechanically loaded materials consists of two different stages, preliminary local re-grouping of valent bonds due to thermal fluctuations and switch-over of oxygen bridge bonds. At first stage, low-activated glass network extension proceed. Due to thermal fluctuations, basic structural elements ($SiO^4$ tetrahedrons) undergo deformation and regrouping of their valent Si-O bonds, i.e. the bond angles slightly change their values and atoms in Si-O-Si bridges become displaced from their previous positions. As the interaction between covalent atoms is short-range and oriented, the pair interaction energy sharply increases at large changes of interatomic distance. The activation energy of atom regrouping proceeding without chemical bond rupture is close to the mean thermal motion energy of atoms in glass melt; and, according to estimates, 100-150 atoms should participate in formation of microvoids of the required size (Nemilov, 1978). Such local fluctuations of valent bond configurations are necessary for subsequent elementary kinetic acts, namely switch-over of oxygen bond in Si-O-Si bridges (Filipovich, 1978; Mueller, 1955, 1960). Here, one of the oxygen bonds with silicon atoms undergoes disruption, and it switches over to another Si-atom having a vacant bond (Fig. 5). As a result of the rearrangement process, the O-atom shifts at one interatomic distance, and that is possible due to the presence of a microvoid earlier formed in the vicinity of the O-atom. Such bonds switches over occur in crystals during dislocations glides in the time of plastic flow.

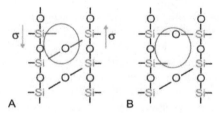

**Figure 5.** Mechanism of viscous flow in $SiO_2$ (or $GeO_2$) glass proceeding as a result of silicon vacant bond switching in Si-O-Si bridges: A - initial state; B - after Si vacant bond switching

The activation energy of the bridge oxygen switch-over is a sum of two components. The first component is the relatively low microvoid formation energy, or fluctuating local silicon-oxygen network deformation energy ~4 - 5 kcal/mol (Sanditov, 1976). Herein, the energy barrier of the second process, direct switch-over of bridge oxygen bonds, decreases to some minimal level of ~20 - 30 kcal/mol, which is different for various glasses. However, if a small number of atoms (3 - 4) participate in the bridge oxygen bond switch-over process, up to ~100-150 atoms are involved in the local deformation of the silicon-oxygen bulk. The latter circumstance provides for a substantial contribution of the entropy term to the effective glass viscosity activation energy, $H^*_\eta$. Vacant bonds switch-over activation energy in the glass viscosity theory is connected with formation mechanism of bonds (Filipovich, 1978; Nemilov, 1978; Sanditov, 1976; Zakis, 1981; Mueller, 1955, 1960; Nemilov, 1978), i.e. with a predominant way of vacant bonds formation and switch-over type. However, rather low activation energies of ~25-35 kcal/mol and large pre-exponent multipliers are typical of the temperature dependencies of bond switch-over rates, providing for most defects migration in the glass network and for the viscous flow of glass at $T \sim T_g$. Therefore, the processes conditioned by directed bonds switch-over (viscous flow, crystallisation, chemical interaction, and diffusion) proceed, as a rule, at relatively low rates, which is typical of the chemical properties of polymeric glass-forming substances (Appen, 1974).

All the above regularities suggest that, in GeO$_2$<Ge-NCs> heterolayers, oxygen vacancies generated in glassy GeO$_2$ matrix at T > 500$^0$C (i.e. at temperatures close to $T_g$(GeO$_2$)) at the boundary of Ge-nanoparticles can easy move within matrix volume. Translation of bridge oxygen bonds to vacant bonds of Ge-atoms underlies migration of oxygen vacancies. Maximum values of equilibrium vacancies concentrations in GeO$_2$ glass are high; as a result, this mechanism becomes capable of providing mass transfer of Ge-atoms through oxide in the heterolayer towards the boundaries of Ge-nanoclusters. In this way this mechanism contributes to Ge-nanoclusters growth. During decay of GeO(solid) atomic network at temperatures over ~250 °C, the decomposition process of this network, leading to releasing of half the total amount of Ge-atoms initially contained in GeO(solid), ensures excessive supply of vacant oxygen bonds towards the forming GeO$_2$ matrix. The rearrangement of valence bonds of Ge-atoms in the sp$^3$-hybridization form is accompanied by the release of considerable energy. During rearrangement, the switch-over of bridge oxygen bonds to vacant bonds requires the least energy in comparison with all other potential barriers. For a number of glasses, the activation energy of the switch-over falls in the range from 23 to 37 kcal/mol, these values are close to the activation energy of oxygen atoms hops in alkaline-silicate glasses (Gorokhov, 2005).

### 3.3. Densification of GeO$_2$<Ge-NCs> heterolayers and them moisture absorption

The majority of physical and chemical processes in the modification of lattices in GeO(solid) and GeO$_2$ layers should be analysed in terms of structural rearrangements with active participation of oxygen vacancies. First of all, such processes include densification of heat-treated layers and adsorption/desorption of moisture from atmosphere in the material.

The densification process is due to the fact that the high rate of GeO molecules condensation from vapour at low temperatures (< 200-250 °C) forms low-density GeO(solid) layers with high concentration of lattice defects. Concentrations of various point defects and their ensembles, pores and other microvoids in the film lattice are much higher than the values typical for the quasi-equilibrium state of atomic network. Such a structure is typical of CVD $SiO_2$ layers, especially deposited at T < 500 °C. This structure has a zeolite-like behaviour, which disappears completely after annealing at T > 700 °C and is manifested in the ability of CVD $SiO_2$ films during the storing to absorb in their volume water molecules and other atmospheric vapours. The absorbed moisture fills all kinds of defects in the volume of the film atomic network. Besides, part of $H_2O$ molecules absorbed in the film dissociates into hydroxyl groups OH- and hydrogen ions $H^+$ fix at vacant bonds of atoms in the oxide lattice.

Annealing of CVD $SiO_2$, GeO(solid) and $GeO_2$ layers at temperatures over 250-300 °C activates processes leading to a decrease of nonequilibrium defects concentration in the atomic lattices of the oxides. Usually this leads to an increase of film density with simultaneous decrease of film volume (shrinkage), as well as to increased optical constants, decreased conductivity, increased tensile stress and viscosity of the material. The intensity of the process and the degree of film properties change depend on the temperature, anneal duration and atmosphere in which anneals were held, as well as on the defects density (or imperfection) of the initial atomic network of the film. Typically, the stronger the thermal treatment and more defective the films, the higher the degree of change of basic film characteristics. For instance, the shrinkage degree in thickness of CVD $SiO_2$ and $Si_xN_y(H)$ films varies within the range from 3-5 to 25-30% (Gorokhov et al., 1982; Pliskin et al., 1965). The presence of absorbed moisture in the film increases its ability to shrink. Therefore, protective (cap) layers are used for protection of GeO(solid) and $GeO_2$ layers from the degrading effects of atmosphere. Thin (tens of nm) low-temperature PE $Si_xN_y(H)$ films are the best encapsulating coatings. The protection ability of CVD $SiO_2$ layers is worse, so one should increase their thickness. Note that during anneals moisture leaves the GeO(solid) and $GeO_2$ layers faster than CVD $SiO_2$ layers. Anneals at 150-180 °C during ~ 30-40 min in vacuum or in a flow of a pure and dry inert gas are sufficient for effective moisture desorption from germanium oxide layer 150-200 nm thickness.

### 3.4. Other methods of structure modification of $GeO_2$<Ge-NCs> heterolayers

The composition and structure of $GeO_2$<Ge-NCs> heterolayers can be easily modified chemically. Removal of the glassy $GeO_2$ matrix from $GeO_2$<Ge-NCs> heterolayer is the simplest way. Glassy $GeO_2$ is readily soluble in water and, especially, in aqueous solutions of HF, whereas crystalline or amorphous Ge-nanoparticles remain unaffected by such solutions. On dissolution of $GeO_2$ matrix, Ge-nanoclusters initially contained in the volume of a $GeO_2$<Ge-NCs> heterolayer agglomerate with each other due to very weak electrostatic forces and settle on the substrate forming a weakly coupled, very loose and porous Ge-layer. So, very thin $GeO_2$<Ge-NCs> heterolayers can be used to produce, for example, single-layer coatings formed by Ge-nanoparticles on substrates of different materials. An annealed agglomerative layer of Ge-nanoparticles becomes sintered, and a more durable Ge-coating with a porous structure forms. Such a layer was formed inside a multilayer structure containing SiO$_2$ films in order to examine the effect of femtosecond (fs) laser treatments on this layer (see Fig. 6 a). Under ablation, after the explosive action of fs laser pulses a region of porous germanium was detached from the thick SiO$_2$ layer, turned over and thrown away onto the undestroyed part of the multilayer system. As a result, we became able to obtain a SEM image of the upper and lower surfaces of the layer under study (see Fig. 6 b).

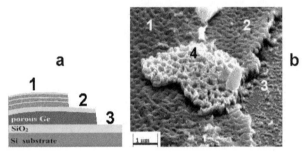

**Figure 6.** Surface morphology of a multilayer coating after fs laser treatment : **a** - scheme of coating structure consisting of three tiers: the bottom tier is Si-substrate with the first SiO$_2$ film (of 100 nm thick), middle tier is a Ge-porous layer (of ~ 70-230 nm thick) coated by second SiO$_2$ film (of 100 nm thick), and upper level is three bilayer structures, each of them consists of thin heterolayer GeO$_2$<Ge-NCs> (of ~ 15 -25 nm thick) coated by the SiO$_2$ cap-layer (of 10 nm thick) (CVD SiO$_2$ is shown as blue, Ge-porous and GeO$_2$<Ge-NCs> layers – as brown); **b** - SEM image of the multilayer coating surface: **1** – unmodified by radiation, **2** – after liftoff of the top tier; **3** - surface of the lower SiO$_2$ layer after liftoff of the Ge-porous layer; **4** - surface morphology of the back side of the Ge-porous layer fragment produced by laser damage.

Processes of oxidation and shrinkage of GeO(solid) layers and GeO$_2$<Ge-NCs> heterolayers are similar. We found that oxidation of GeO$_2$<Ge-NCs> heterolayers in an oxygen flow at normal pressure differs little from the oxidation of pure germanium wafers under the same conditions. The process can be well monitored ellipsometrically at temperatures over ~500$^0$C. Under such conditions, GeO(solid) layers disproportionate into Ge and GeO$_2$ according to reaction (3) for a time not longer than 2-3 minutes, and then they undergo oxidation as GeO$_2$<Ge-NCs> heterolayers. Ge-nanoparticles oxidation front propagates from the surface of the film into depth. The kinetics is close to kinetics of pure germanium oxidation even quantitatively. The mechanism of the process is that the Ge/GeO$_2$ interface in the system generates vacant bonds ("oxygen semi-vacancies") on Ge-atoms incorporated in the formed oxide network. Further on, these vacant bonds tend to maximally fill the GeO$_2$ lattice by their translation mechanism. O-atoms from the molecules of ambient O$_2$ get captured by vacancies on the exposed external surface of the oxide layer. After this captured O-atom becomes attached with one of its bonds to the GeO$_2$ network while the other bond usually remains free. In the oxide this free bond presents one quarter of a Ge-atom vacancy and is excessive (non-equilibrium). Thus, the interdiffusion of oxygen semi-vacancies and quarters of germanium vacancies in the oxide layer network, accompanied by gradual annihilation of these complementary defects, is the oxidation process of both germanium wafers and Ge-nanoparticles in the glassy $GeO_2$ layer.

Crystallisation of the glassy $GeO_2$ matrix in $GeO_2$<Ge-NCs> heterolayers is an exceptionally structural (not chemical) process. We have studied this process using layers of pure thermal germanium oxide on Ge(111) substrates. The most important is that $GeO_2$ layers begin to rearrange into the low-quartz structure already at 600-620 °C; it is worth noting that there are conditions for forming large-area continuous block-epitaxial coatings from crystallised hex-$GeO_2$ over quite a large area. Interestingly, an analogous process also proceeds in thermal silicon oxide with formation of block-epitaxial β-crystabalite $SiO_2$ layers. But the latter process proceeds very slowly even at temperatures T > 1200°C. It makes the material practically unpromising in the silicon integral planar technology. Unlike $SiO_2$, low-quartz $GeO_2$ layers are more suitable for application, as crystallisation of glassy $GeO_2$ layers considerably increases the mechanical strength and chemical stability of the films. For removing crystallised layers, etching solutions with considerable HF acid proportion are necessary. It is described in the monography (Knoss, 2008) where reactions of glassy $GeO_2$ layers with CVD $SiO_2$ and $Si_3N_4$ films deposited onto such layers were also described.

Interaction of thermal $GeO_2$ layers with ammonia at ~ 700-750°C according to reaction $3GeO_2 + 4NH_3 \rightarrow Ge_3N_4 + 6H_2O\uparrow$ proceeds in the oxide lattice in compliance with the micromechanisms similar to those involved in germanium oxidation (Gorokhov, 2005). At these temperatures the solid reaction product, $Ge_3N_4$, dissolves in the practically liquid oxide layer and forms a glassy $GeO_2:Ge_3N_4$ (m:n) layer. Under the described conditions, germanium (both the wafer material and Ge-nanoparticles) also reacts with $NH_3$ and forms $Ge_3N_4$. Therefore, relatively thin $GeO_2$<Ge-NCs> heterolayers can be used to increase the $Ge_3N_4$ content of the formed $GeO_2:Ge_3N_4$ glassy layers.

The above reactions are interesting because they allow a drastic modification of the structure and composition of the $GeO_2$ matrix in $GeO_2$<Ge-NCs> heterolayers up to ~100 nm thickness. Low viscosity of glassy $GeO_2$ at temperatures T>Tg provides for rapid mutual solution of $SiO_2$ layers with germanium dioxide, practically to a homogeneous composition at thickness amounting to hundreds of nanometers; this process transforms the layers into germanium-silicate glass. Varying the thickness ratio of $GeO_2/SiO_2$ double-layers we can change the composition of composite $GeO_2:SiO_2$ glass in a wide range of values.

Glassy $GeO_2$ reacts with CVD $Si_3N_4$ films at the interface at T > 700°C according to the formula $3GeO_2 + 2Si_3N_4 \rightarrow 2Ge_3N_4 + 3SiO_2$. Like $SiO_2$, the low viscosity of glassy $GeO_2$ alleviates the interdiffusion of initial and final products of this reaction across the interface of the reacting layers. As a result, a layer of glassy $GeO_2:Ge_3N_4:SiO_2$-type compound several tens-hundreds of nm thickness forms at the interface. The composition of this layer depends on the thickness ratio of initial layers, and on the temperature and duration of the reaction. Under heating held at temperature T ~750°C for an hour, in the obtained germanium-silicate glass and multi-component $GeO_2:Ge_3N_4:SiO_2 \rightarrow$ k:p:q layers crystallisation begins, proceeding similar to that in pure glassy GeO$_2$. The crystal lattices of the glasses are isomorphic to the low-quartz structure, and they excel hex-GeO$_2$ layers in chemical stability.

Interestingly, the dramatic increase of density and chemical stability in composite glassy GeO$_2$:SiO$_2$ and GeO$_2$:SiO$_2$:Ge$_3$N$_4$ – k:p:q layers begins already at their formation stage. With the example of GeO$_2$:SiO$_2$, it is seen that this effect is much more pronounced in comparison with the effect due to mere linear growth of the more chemically stable SiO$_2$ fraction in the two-component composition. In other words, the etching rate of the glassy GeO$_2$:SiO$_2$ layer in a structurally sensitive etchant turns out to be a few times lower than the individual etching rates of glassy GeO$_2$ and SiO$_2$ layers (even when the fraction of SiO$_2$ in the composition is not high). We suppose this effect been due to the fact that the formation of glassy GeO$_2$:SiO$_2$ and GeO$_2$:Ge$_3$N$_4$:SiO$_2$ – k:p:q films occurs in the vicinity of the crystallisation temperatures of these substances. Therefore, during annealing such glasses should undergo the stage preceding crystallisation. At this stage, the glass atomic network is getting ready to transform into a high-ordered crystal lattice so that the density of quasi-equilibrium defects rapidly decreases. Crystallised glass is known to have lattice defect concentrations orders lower than the initial concentrations of the same defects at the initial stage. Hence, introduction of GeO$_2$ into glassy SiO$_2$ provides for a sharp decrease of vitrification and crystallisation temperatures of the composition, together with a sharp decrease of vacancy concentration. Thus, CVD SiO$_2$ layers with reduced density, due to a high concentration of non-equilibrium defects in rather unstable atomic network, can be etched in etchant sensitive to their structure tens of times faster than quasi-equilibrium thermal SiO$_2$ layers formed at temperatures ~ 1000 $^0$C (Pliskin and Lehman, 1965).

Thus, the described structural-chemical modification processes of glassy oxide GeO(solid), GeO$_2$ layers and GeO$_2$<Ge-NCs> heterolayers have a high modification ability, which allows us to obtain chemically stable coatings with an original stable structure and chemical composition, high mechanical strength and easily controlled physical, electrical and optical properties from chemically unstable, structurally imperfect, physically and mechanically soft, and electrically unstable layered films.

## 4. Laser treatments of GeO$_2$<Ge-NCs> heterolayers

Thermal anneals, traditionally used in microelectronics, were the tool for the described modifications of the composition and structure of the films under study, in which different Ge oxide layers played the main role. As it was noted in Introduction, pulsed laser irradiations offer a considerable resource for enlarging the list of possible chemical and structural transformations can be implemented in the materials of interest. Irradiation of film systems with electromagnetic radiation can be considered as a basic kind of possible treatments. Unlike common thermal treatments, irradiation of samples with light pulses allows a radical widening of the range of conditions that can be realized in the high-energy treatment of layered materials. First, here we mean the value of the energy input in irradiated materials, which can be orders higher than the limits that can be reached in ordinary heat treatments of samples. Second, the range of dynamic parameters of high-energy effects in irradiated films can also be widened by many orders, together with the spatial localization parameters of such effects. Due to these factors, irradiation of samples with short laser pulses considerably extends the spectrum of possible modifications of thin films in micro- and nanoelectronic structures.

Over more than thirty years, Pulsed laser Annealing (PLA) has been successfully used to achieve crystallisation and re-crystallisation of semiconductor materials in thin-film structures (Dvurechenskiy et al., 1982). Earlier, PLA treatments of thin-film structures with light quanta exceeding in energy the bandgap width of semiconductor material were recognized as a useful means for rapid (typically, during some tens of nanoseconds or less) restoration of the crystalline structure in disordered or even completely amorphized near-surface layers of Si (Ahmanov et al., 1985). At a proper choice of laser parameters, almost all laser radiation can be absorbed within the film; hence, this radiation does not reach the wafer and does not heat it. As a result, short laser pulses allow one to avoid overheating of the substrate during film cooling due to diffusion of heat into the substrate (tens of nanoseconds). That is why PLAs can be used to crystallize amorphous silicon films on substrates not withstanding high temperatures.

Powerful laser impacts lead to fast, high-quality re-crystallisation of amorphized near-surface semiconductor layers. Despite the fact that PLA has already become a well mastered technique, laser annealing experiments raised a number of still unsolved fundamental physical questions (Chong et. al., 2010). Recent contributions have shown that fast laser-induced phase transformations in near-surface semiconductor layers, such as melting-hardening, phase transitions 'amorphous solid – crystal' and 'crystal – amorphous solid', etc., proceed over nano-, pico-, and even subpicosecond time scale. To explain the various and, in many respects, unexpected phenomena observed during PLA treatments of semiconductor structures, it is necessary to give an answer to a number of fundamental questions regarding the behavior of semiconductors in strong laser fields.

PLA is widely used to eliminate structural imperfections and radiation defects introduced during implantation of ions in the near-surface crystal layers (Kachurin et al., 1975). PLAs provide an important possibility for the technology, obtaining perfect crystal structures in subsurface layers with impurity concentrations unreachable in the common thermal annealing (to $10^{21}$ cm$^{-3}$ and higher). Diffusion of impurities is highly suppressed under PLA conditions (Kachurin et al., 1975). In addition, PLA allows realization of processes that do not occur during common thermal treatments. Such processes, which are of physical concern, are conditioned by the presence of dense plasma in the samples. The range of observed PLA-induced processes includes softening of phonon modes with increasing plasma density, change of bandgap width with increasing $n_e$ (non-equilibrium carrier concentration), registration of optical and electrical phenomena in semiconductors depending on $n_e$ (Ahmanov et al., 1985), etc. Spatial coherence of laser radiation used for pulse laser treatments of semiconductor materials allows creation on the surface of annealed samples of periodic spatial structures, e.g. lattices formed by alternating crystalline and amorphous regions (the so-called interference laser annealing).

Later, nanosecond pulsed treatments using excimer XeCl ($\lambda$ = 308 nm) (Volodin et al., 1998) and ArF laser radiation ($\lambda$ = 193 nm) were used to obtain silicon nanoclusters and achieve crystallisation of a-Si inclusions in SiN$_x$ and SiO$_x$ films (Rochet et al., 1988). Some of the unusual properties of such heterolayers were manifested in samples treated with ultrashort pulsed fs laser radiation (Korchagina et al., 2012). Reports on laser treatments performed to modify the structure, chemical composition, and properties of GeO(solid) and GeO$_2$ films, and also to achieve formation of c-Ge nanoparticles and crystallisation of a-Ge inclusions in dielectric matrices are encountered much more rarely (Gorokhov et al., 2011).

## 4.1. Peculiarities of conventional nanosecond PLA

Considering the complexities of PLA impact mechanisms on thin-film semiconductor structures and a complex nature of physicochemistry and structure of glassy GeO(solid), GeO$_x$, and GeO$_2$ films, and GeO$_2$<Ge-NCs> heterolayers, unusual effects could be expected in our films during PLA treatments. Two lasers with different photon energies and durations of light pulses, a KrF excimer laser (wavelength $\lambda$ = 248 nm, pulse duration 25 ns) and a Ti-Sapphire laser (central wavelength $\lambda$ = 800 nm, pulse duration < 30 fs), were used in this investigation. As expected, the impacts of short-pulse treatments on our samples for the two lasers turned out different.

The effect of ns radiation was studied for two bilayer systems, GeOx/SiO$_2$ and Si/GeO$_2$<Ge-NCs>/SiN$_x$O$_y$, both prepared on Si(100) substrates (SiO$_2$ films of thickness 100 nm and SiN$_x$O$_y$ films of thickness 25 nm were the cap layers). A sketch of the second system is presented in Fig. 7 a, b. The laser beam was focused, through a square-section diaphragm (200 x 200 µm), on the sample surface. Laser pulses followed at a frequency of 100 Hz, and the beam moved along one of the sides of the square on the sample in ~180-µm increments. After exposure of one band on the sample surface to the beam with a set laser pulse energy (see Fig. 7 d) the beam was shifted a step further to treat the next band under the first one. An exposed film section was formed out of several bands. Because of overlapping of laser spots, the film was exposed to the laser beam two times on the edges of the square and four times in the corners (Fig. 7 c, d and e). Such neighborhood of film regions exposed to different numbers of identical laser pulses (1, 2 and 4) allowed us to trace the pulse-by-pulse changes induced in the regions where the film system acquired different thicknesses.

In the absence of cap layers, the used PLA completely "evaporated" the germanium oxide layers on our samples transforming them into GeO(gas) according to reaction (4). The SiO$_2$ and SiN$_x$O$_y$ cap layers, impeding the removal of GeO(gas) molecules, substantially slowed down the formation reaction of those molecules in the protected films. Nonetheless, even in the presence of cap layers the irradiated regions exhibited a notable change in interference colors (Fig. 7 b, c, d, e).

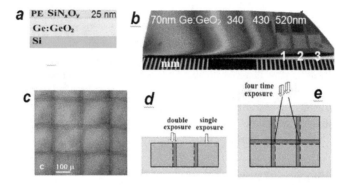

**Figure 7.** **a** – schematic representation and **b** - photo of the Si/GeO₂<Ge-NCs>/SiNₓOᵧ bilayer system on Si substrate irradiated with laser radiation to fluences 170, 150 and 130 mJ/cm² in regions 1, 2 and 3 of the heterolayer (thicknesses ~390, ~480 and ~520 nm, respectively); **c** - micrograph and **d, e** – schematic illustrating the formation of single-, double- and four-time exposed areas of the film at stepped irradiation.

### 4.1.1. Shrinkage of GeO₂<Ge-NCs>/SiNₓOᵧ bilayer coatings under ns PLA

Data obtained with a set of experimentaltechniques in irradiated regions 1, 2 and 3 of the sample with ~390-, ~480- and ~520-nm thick heterolayers provided a comparison of the type and degree of changes in the structure and material properties of examined films after their exposure to different laser fluences, 130, 150, and 170 mJ/cm². Changes of interference colors in the bilayer system occurred due to strong shrinkage of the materials forming in the system individual layers, which was usually accompanied by increase of film optical constants. The effect here is similar to the one which was studied by ellipsometry and which proceeded with a change of optical constants of GeO(solid) layers during thermal anneals. In the latter case, GeO(solid) did not evaporate because of low annealing temperature. Results of ellipsometric measurements of our samples performed following their thermal treatments in vacuum are shown in Fig. 8. The measurements were carried out using a special method that combined spectral and scanning ellipsometry with multi-thickness measurements (Marin et al., 2009) and that allowed us to obtain sufficiently precise data on optical constants of annealed GeO₂<Ge-NCs> heterolayers throughout the whole visible range of the spectrum.

The data for $k(\lambda)$ (curves 1 - 4 in Fig. 8) show the changes in the bandgap of as-deposited GeO(solid) films and GeO(solid) films given traditional anneals at different temperatures. These changes are the result of the chemical decomposition of film material connected with the appearance in the film and subsequent growth of Ge nanoparticles in which, due 3D confinement effects, the efficient band gap value decreases with an increase of nanoparticle size, tending at large nanoparticle sizes to the bandgap value of bulk $\alpha$-Ge. The $k(\lambda)$ data for all examined films were plotted in the Tauc coordinates, with the square root of absorbance coefficient $\alpha$ being plotted along the vertical axis. This coefficient was obtained from the extinction coefficient by formula $\alpha_j(E)=2\pi k_j(E)/\lambda$; here, the parameter $j$ is the number of a studied film in Fig. 8. Using the well-known formula for the absorption coefficient of indirect-band semiconductors, $\alpha^{\frac{1}{2}} \sim (E - E_g)$, and interpolation of the linear dependences of $\alpha^{\frac{1}{2}} (E- E_g)$ to $(E- E_g) \rightarrow 0$, we were able to evaluate the effective optical gap $E^j_g{}^{eff}$ in the GeO(solid) films and in the GeO$_2$<Ge-NCs> heterolayers formed upon decomposition of the films. The variation of the effective optical gap in GeO(solid) films with the growth of annealing temperature T$_{an}$ is shown in Fig. 9. For comparative analysis of obtained $E^j_g{}^{eff}$-values and for analysis of Ge-nanocluster sizes, the values of $E^j_g{}^{eff}$ for amorphous bulk germanium and for GeO$_2$<Ge-NCs> heterolayers are also shown in Fig. 9, curves 5 and 6. For the effective optical gap $E^5_g{}^{eff}$ in GeO$_2$<Ge-NCs> heterolayers, a value of 1,15 eV was found, and the mean size of Ge NCs in the heterolayers proved to be ~5 -6 nm (according to TEM data). Since the value of $E^j_g{}^{eff}$ in annealed GeO(solid) films always remained appreciably higher than 1,15 eV, it can be concluded that a-Ge nanoparticles in those films were smaller than nanocrystal sizes in GeO$_2$<Ge-NCs> heterolayers. These data have also allowed us to evaluate the sizes of a-Ge nanoparticles in other examined films.

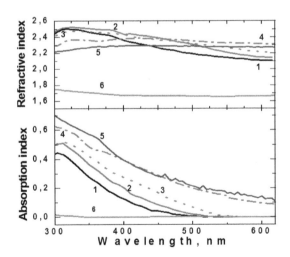

**Figure 8.** Spectral dependences of optical constant (n($\lambda$) and k($\lambda$)) in the various films: 1, 2, 3, 4 – as-deposited GeO(solid) film and films annealed at 260, 290, and 320 $^\circ$C, 5 – LP CVD GeO$_2$<Ge-NC> heterolayer, 6 - thermal GeO$_2$ film grown on a single-crystal Ge(111) substrate.

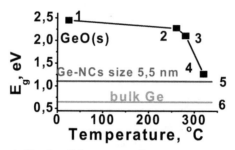

**Figure 9.** The effective optical bandgap $E^{eff}_g$ versus annealing temperature: **1** – in as-deposited GeO(solid) films (with Ge-NCs sized 0,2 - 0,4 nm) and after subsequent anneals (**2** - 260 °C, 6 min; **3** - 290°C, 4 min, **4** – 320°C, 4 min); **5** – $E^{eff}_g$ for a GeO$_2$<Ge-NCs> heterolayer with 5,5-nm Ge NCs; **6** - $E^{eff}_g$ for bulk Ge.

According to AFM data, on a sample area irradiated with one laser pulse the shrinkage degree of the film was ~5 - 9%; this value increased up to ~23 - 34% under four laser pulses (Figs. 10 and 11). The cause for the film shrinkage was the low initial density of the film conditioned by the presence of many defects in the atomic network of GeO(solid). Similar data were also obtained for GeO$_x$ layers covered by a cap SiO$_2$ layer (on a silicon substrate) in which the predominant part of the GeO(solid) component has already decayed during the growth of those layers. Notable film shrinkage is usually observed on anneals in loose films grown by CVD at low temperatures. Normally, thermal treatments modify the structure of CVD films and improve their quality. Anneals of multilayer structures activate both the shrinkage processes in individual layers of multilayer films and the chemical reactions proceeding among neighboring layers (Knoss, 2008). The high defect content alleviates mass transfer processes in the layers and rearrangement of their structure.

### 4.1.2. Changes in the structure and properties of thin bilayer coatings under ns PLA

Raman spectroscopy and microprobe IR-spectroscopy were used to analyze the structure of germanium inclusions in examined films. The non-destructive Raman method combined with calculations is a very informative tool for nano-object studies. The presence of Raman peaks at 301,5 and 520 cm$^{-1}$ in the spectra is an indication of c-Si and c-Ge micro-inclusions present in the layers in which light is scattered by long-wave lattice phonons. It is in this way that micro- and nanoparticles are usually detected in the films of these materials. By the way, if the crystals are small-sized (~nm), then the spectral position of related Raman peaks exhibits a red shift in comparison with the spectral position of the peaks due to bulk c-Si and c-Ge. According to the peak shift value, one can determine the mean size of semiconductor nanocrystals in the films of interest using the phonon localization method (Volodin et al., 2005). The greater is the Raman shift of the peaks toward smaller wavenumbers in comparison with the position of the peaks in bulk c-Ge (301,5 cm$^{-1}$), the smaller is the typical size of Ge nanocrystals in the films (Fig. 12). Broad scattering bands peaking at 275 - 280 cm$^{-1}$, which are due to the presence of amorphous germanium, are also clearly seen in the Raman spectra of examined GeO(solid) films and GeO$_2$<Ge-NCs> heterolayers since it is into this spectral region where the maximum density-of-state value of optical vibrations in a-Ge falls. In the films under study, amorphous germanium is normally contained in the form of nanoparticles dispersed in glassy $GeO_2$ matrix (Knoss, 2008).

The decrease of film thickness and increase of film material density after PLA as manifested in the Raman spectra of $Si/GeO_2<Ge\text{-}NCs>/SiN_xO_y$ films is accompanied with the growth of the intensity of the peak at 520 cm⁻¹ due to light scattering by long-wave optical phonons

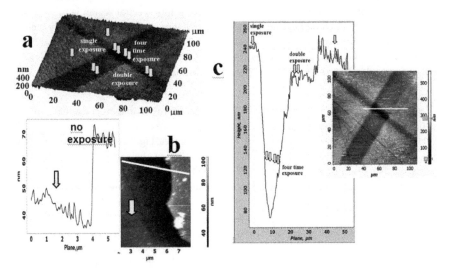

Figure 10. a - AFM image illustrating the shrinkage of the $GeO_2<Ge\text{-}NCs>/SiN_xO_y$ bilayer system in region 2 (see Fig. 7 b) after PLA of the sample with ns laser pulses (fluence $E_0$=150 mJ/cm²); b - measurements of the shrinkage of the bilayer film in the region treated with one laser pulse, c – the same for two and four laser pulses. The arrows ↓, ↓↓ and ↓↓↓↓ indicate regions subjected to the single, double and four-time exposures.

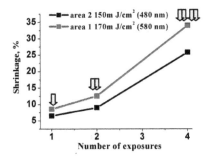

Figure 11. Effect of the number of laser pulses, laser fluence, and thickness of $GeO_2<Ge\text{-}NCs>$ heterolayer on the shrinkage value of the $Si/GeO_2<Ge\text{-}NCs>/SiN_xO_y$ bilayer system.

in the Si substrate (Fig. 12). This effect is manifested as an increase of the transparency of the bilayer film for the excitation laser radiation in Raman measurements (514,5 nm Ar+ laser) falling onto, and reflected by, the substrate. The cause for the shrinkage effect are PLA-induced changes of optical constants of the films in the $Si/GeO_2<Ge-NCs>/SiN_xO_y$ bilayer system, as optical constants of Si substrate remained essentially unaltered in the spectral region around the excitation wavelength. Let us consider now possible reasons for such changes. Some minor increase in the refractive index of a very thin $SiN_xO_y$ film due to its shrinkage and compaction cannot be the factor appreciably affecting the optical constants of the bilayer system. The absorption index of the film in this spectral region is initially very small, and the irradiation of the sample with laser radiation could not considerably increase its value. Hence, the increase in transparency of the bilayer system was most probably due to the change in the optical constants of the $GeO_2<Ge-NCs>$ heterolayer or, alternatively, it could be a result of the chemical interaction of the two layers in the interfacial region of the structure.

Indeed, micro-Raman and micro-IR-spectroscopy data showed that such processes indeed proceed in the film system under study. Raman spectra taken from local areas of the film system irradiated with one, two, or four laser pulses (see Fig. 12 a) were measured with excitation laser beam focused to a ~1-μm diameter spot on the sample surface. From the registered changes in Raman spectra, we were able to judge the rate of release of Ge atoms during decomposition of the GeO(solid) structure, and also the formation of amorphous ($\alpha$-Ge) and, then, crystalline (c-Ge) phases of Ge. The formation kinetics of $\alpha$-Ge and c-Ge phases within the heterolayer is characterized by initially the highest amorphous germanium content of the layer system and by almost complete absence of c-Ge phase from it (see the initial spectrum in Fig. 12 a). After one laser pulse, the fraction due to $\alpha$-Ge component sharply decreased in the system, and c-Ge phase emerged in quite large quantities. Two and four laser pulses gave rise to an increased amount of c-Ge phase at a comparatively slow regeneration of $\alpha$-Ge component. During annealing in these

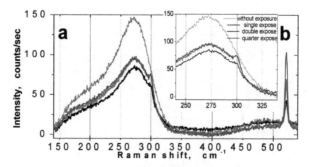

**Figure 12. a** - Raman spectra illustrating the influence of the number of ns laser pulses given to a local region on the sample (region 2 in Fig.7 b, $E_0$ =150 mJ/cm²) on the structure of the $GeO_2<Ge-NCs>$ heterolayer in the $Si/GeO_2<Ge-NCs>/SiN_xO_y$ bilayer system; **b** – magnified fragment of the same spectra.

heterolayers, amorphous germanium phase usually forms a big amount of amorphous Ge nanoparticles growing in size during high-energy treatments of the system, and then such nanoparticles undergo crystallisation without interrupting their continuous growth. The growth of Ge nanoparticles proceeds primarily due to decay of nanocluster fractions with minimal sizes. The formation processes of $\alpha$-Ge and c-Ge nanoparticles are also maintained by the decay of the atomic network of metastable GeO(solid), which supplies the nanoparticles with atomic germanium.

According to micro-IR-spectroscopy data (Fig. 13), the atomic network of the insulator matrix also undergoes modifications that proceed simultaneously with the transformation processes of germanium inclusions in the GeO₂<Ge-NCs> heterostructure. Yet, the process turned out different than the expected formation of pure glassy GeO₂ by reaction (3) during anneals of GeO(solid) films (Fig. 4 b). In the heterolayer region treated with one laser pulse the IR absorption band starts moving towards longer wavelengths instead of showing the expected trend to the maximum at 860-870 cm⁻¹. On the other hand, the maximum IR absorption intensity in the region treated with two laser pulses decreases markedly, and the whole band notably widens. In the bilayer structure under study, in addition to the formation of GeO₂ during decomposition of metastable GeO(solid), the GeO₂<Ge-NCs> heterolayer and the SiNₓOᵧ film can react with each other in the interfacial region. The GeO₂ matrix of the heterolayer, and the Si₃N₄ and SiO₂ materials forming the cap SiNₓOᵧ layer, are involved in this reaction. In particular, the products of the reaction between GeO₂ and Si₃N₄ proceeding at T>650⁰C ($3GeO_2 + Si_3N_4 \rightarrow 3SiO_2+Ge_3N_4$), SiO₂ and Ge₃N₄, can be dissolved by glassy GeO₂ matrix. As described above, these processes finally yield a GeO₂:SiO₂:Ge₃N₄→ k:p:q glassy compound. Since pure Ge₃N₄ and glassy SiO₂ exhibit IR absorption peaks respectively at about 770 cm⁻¹ and 1070 cm⁻¹, it can be expected that the peaks due to the Ge₃N₄:GeO₂ and GeO₂:SiO₂ double glasses will tend to shift in opposite directions from the peak due to pure GeO₂; i.e. the first peak will shift towards longer wavelengths and the second peak, towards shorter wavelengths. On the whole, the material of the GeO₂ matrix is expended on the formation of both SiO₂ and Ge₃N₄, and also on the formation of the three-component GeO₂:SiO₂:Ge₃N₄ → k:p:q compound. As a result, a predominant part of the cap SiNₓOᵧ layer may turn into Ge₃N₄ and SiO₂ that will subsequently undergo dissolution in the upper part of the GeO₂<Ge-NCs> heterolayer. Such transformation will be capable of reaching depths comparable with the thickness of the initial cap layer. This may become the reason for the observed modification of the IR absorption band in the irradiated film, including reduced absorption at the band maximum, shift of the band, and its strong widening (Fig. 13).

Of course, the properties of the upper layer in the initial bilayer structure will undergo dramatic changes. In particular, the optical constants and the optical bandgap will strongly change in this layer. Dissolution of wide-bandgap glassy SiO₂ ($E_g^{SIO2} \sim 9$ eV) in the matrix of the GeO₂<Ge-NCs> heterolayer will radically change the energy-band characteristics of the heterosystem. A similar behavior is also demonstrated by glass-like Ge₃N₄, which is close to GeO₂ in terms of optical gap, $E_g^{Ge3N4} \sim 5\text{-}5,5$ eV, yet differs from GeO₂ in its high refractive index (1,90-2,00) and permittivity (~9-11). On the contrary, the refractive index of SiO₂ (1,45) is lower than that of glassy GeO₂ (1,61-1,63) (Gorokhov, 2005).

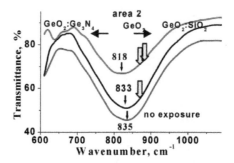

**Figure 13.** Micro IR transmittance spectra illustrating the influence of the number of ns laser pulses given to a local region on the sample (region 2 in Fig.7b, $E_0$ =150 mJ/cm$^2$) on the structure of the dielectric in the Si/GeO$_2$<Ge-NCs>/SiN$_x$O$_y$ bilayer system.

The degree of compositional changes in the upper part of the layer will depend on material viscosity. In the described process, this degree will grow from one to the other irradiation pulse as regions with altered chemical composition will spread away from the interface across both films of the bilayer system. In turn, the compaction of the material in each layer will more and more impede the diffusion process in the intermixing region of the bifilm. Formation of fine cracks at the periphery of the squares irradiated with laser pulses serves an indication for the growth of viscosity in the upper part of the bilayer structure. The formation process of such cracks is defined by two factors: (i) growth of the internal tensile stress in the film due to material compaction (the shrinkage of the film in the direction normal to the film plane does not completely eliminate the lateral internal tensile stress in the film) and (ii) growth of viscosity in the compacted layer due to increase of strength characteristics of the material proceeding with simultaneous vanishing of defects from the atomic network and with modification of the chemical composition of the layer (Gorokhov et al., 1998). As a result, quite a thick intermediate layer (up to ~50-80 nm), in which the insulator matrix has a variable chemical composition and altered structural, mechanical, physical and other material properties, forms during the reactions at the interface between the GeO$_2$<Ge-NCs> heterolayer and the SiN$_x$O$_y$ film. There are no Ge nanoparticles in the upper part of this layer, but the concentration of such nanoparticles gradually increases as we approach the pure GeO$_2$<Ge-NCs> heterolayer. The optical constants of the upper layer, as well as its optical gap value, also vary in a complex manner over the thickness of the layer. But, since glassy SiO$_2$, a very broad-band material, presents one of the composite-glass components, the optical gap in the mixture will be everywhere wider than the optical gap in the regions where only narrow-band glassy components, having an optical gap not wider than 5 - 5,5 eV, are present. Hence, in the course of PLA treatments of the samples with KrF excimer laser pulses the upper multi-component layer will become increasingly more transparent to the excitation beam of the Ar+ laser (514,5 nm) in comparison with the bulk of the lower GeO$_2$<Ge-NCs> heterolayer.

Note that all the discussed processes can be realized using traditional heat treatments at $T \geq$ 400⁰C. Yet, the duration of such heat treatments is typically greater than one minute, whereas, in our case, many atoms simultaneously participated in several structural-chemical rearrangement processes for only several tens of ns. To gain an insight into the whole picture of involved processes, let us focus on their most important features. First, AFM data on the shrinkage of the layered structure and IR-spectroscopy data on the modifications of its chemical composition reveal a gradual increase in the rate of structural-chemical transformations in the system from the first to last laser pulse (Fig. 11). Second, the dynamics of irradiation-induced changes in the optical properties of the bilayer film shows a distinct saturation. Those changes first occur rapidly and, then, their rate sharply decreases. The latter is evident from changes in the Raman spectra of the Si/GeO₂<Ge-NCs>/SiN$_x$O$_y$ system observed 275 to 300 cm$^{-1}$ (Fig. 12 a) at the beginning and at the end ot its irradiation with a sequence of ns laser pulses. During the first pulse, the input energy was absorbed more readily by the bilayer film in comparison with subsequent pulses since the changes in the Raman spectra due to the first pulse were manifested more distinctly in comparison with the changes induced by the second pulse and, the more so, by the two subsequent pulse. The successive growth of the intensity of the Raman peak (520 cm$^{-1}$) due light scattering by the lattice of Si substrate for Ar+ laser excitation (Fig. 12) also agrees with the general tendency in pulse-by-pulse modification of film-structure properties although, here, a more steady growth of peak intensity is observed. The effect is explained by an increase of bilayer film transparency, or reduced absorption of laser radiation by the film in this spectral region). A common feature for all pulses was that a predominant part of their energy dissipating in GeO₂<Ge-NCs> was spent on rearrangement of its lattice structure that proceeded  mainly during the time intervals between the pulses.

An unusual thing here is that, with each subsequent radiation pulse, the Si/GeO₂<Ge-NCs>/SiN$_x$O$_y$ hetrosystem becomes more and more transparent to the probing Raman-spectrometer radiation. Initially, we expected quite a contrary thing, namely, an increase in the absorption of Ar⁺ laser beam radiation by the GeO₂<Ge-NCs> heterolayer. Such an increase was expected to be a result of the considerable amount of free Ge atoms released in the heterolayer bulk. Ge atoms become the material for nucleation and growth of many Ge-nanoparticles, which was to be manifested as an increased absorption coefficient of the heterolayer in the visible spectral region. For instance, such a growth of the absorption coefficient was observed during successive anneals of GeO₂<Ge-NCs> heterolayers and GeO(solid) and GeO$_x$ films (Fig. 8). Similarly, during growth of silicon nanoparticles in the bulk of non-stochiometric SiO$_x$ and Si$_x$N$_y$(H) films under chemical and laser treatments no enhanced transparency of such films was observed in the visible spectral region (Korchagina et al., 2012; Rinnert et al., 2001; Volodin et al., 2010). However, it proved to be a very difficult task to reveal the true nature of such an unusual effect in the experiments with the bilayer system under study, as in PLA-treated samples several different chemical and structural processes proceed simultaneously in the bulk of the GeO₂<Ge-NCs> heterolayer and at its boundaries. It was required to simplify the complex of such structural and compositional modifications in the films. Therefore, we first undertook an analysis of material transformations proceeding during traditional heat treatments in a GeO(solid) film on Si substrate without cap layer (ISP SB RAS) and in a $GeO_x$ film on Si substrate protected by a cap layer formed from sputtered $SiO_2$ (~100 nm) (Nancy University, France).

A GeO(solid) layer of uniform thickness 61 nm was characterized with the follows values of optical constants: n=1,86 and k=0 (obtained from many-angle ellipsometric measurements using an He-Ne laser with $\lambda$=632,8 nm at beam incidence angles $45^0$, $50^0$, $55^0$, $60^0$, $65^0$, and $70^0$). A broad band peaking at 280 $cm^{-1}$ due to Ge nanoparticles was observed in the Raman spectra of the as-deposited film (Fig. 14). The film was thin and its decomposition just began; therefore the amplitude of this Raman band was very low, especially at the sample edges (spectrum 3) where the film was less heated by the evaporator during the film deposition process. The highest degree of decomposition of the GeO(solid) film was observed at the centre of the sample (spectrum 2). Note that 4-minute exposition of a local sample area to the excitation laser beam focused to a ~1-$\mu$m diameter spot on the film surface at the centre of the sample caused a strong reduction in the beam scattering by a-Ge nanoparticles (spectrum 4). This effect was due to photo-stimulated oxidation of the particles by air oxygen in the locally heated surface area. Here, the concentration of a-Ge nanoparticles decreased in the film; yet, the film thickness grew in value due to the formation of an additional amount of germanium dioxide in the film. A comparison of the Raman spectrum of Si substrate without a film (spectrum 1) with the other spectra in Fig. 14 shows that deposition of a GeO(solid) layer onto the film surface (spectrum 3), growth of the fraction of a-Ge in the film in the form of nanoparticles (spectrum 2), and transformation of part of a-Ge nanoparticles into glassy $GeO_2$, i.e. increase in the volume of heterolayer matrix (spectrum 4) lead to multiply decreased amplitude of the peak at 520 $cm^{-1}$ in the Raman spectrum of the Si/GeO(solid) system, this peak being due to light scattering by long-wave optical phonons in the Si substrate. The latter finding can be attributed to enhanced absorption of $Ar^+$ laser radiation in the GeO(solid) film for all these cases.

The cap layer prevents evaporation of the $GeO_x$ layer in $Si/GeO_x/SiO_2$ structures (with layer thicknesses 100 nm/100 nm) annealed at temperatures $T>450^0C$. As a result, the initial metastable atomic network of GeO(solid) in such structures undergoes decomposition with the emission of all the excessive germanium and its accumulation in a-Ga nanoparticles. On increasing the annealing temperature, such a-Ga nanoparticles transform in Ge nanocrystals. These processes can be easily monitored using Raman scattering measurements (see Fig. 15). The temperatures used to anneal the $Si/GeO_x/SiO_2$ sample were insufficient for the onset of a reaction between the $GeO_2$ matrix in the decomposed $GeO_x$ film and the $SiO_2$ cap layer; this sample was found to also not absorb the Ar+ laser radiation and undergo no heating due to the wide optical gap of the material. Hence, all the changes proceeding with the Raman scattering peak due to light scattering by phonons in Si substrate bear no relation with modification processes in the structure of the $GeO_x$ layer. On the other hand, it is seen that the process of germanium release in the $GeO_x$ film proceeding with the formation of a-Ge nanoparticles and their subsequent crystallisation successively reduces the amplitude of the peak due to Raman scattering in Si substrate. The same result was also observed in deposition of the initial $GeO_x$ layer, like in the case of GeO(solid) layer synthesis (see Fig.14).

Note that the spectral position of the Raman peak due to Ge-nanocrystals in the $GeO_x$ film is no difference from the spectral position of the peak in bulk Ge crystals (301 cm$^{-1}$). The latter finding is indicative of rather large mean sizes of Ge nanocrystals (~8-10 nm and greater) formed in this film during anneals.

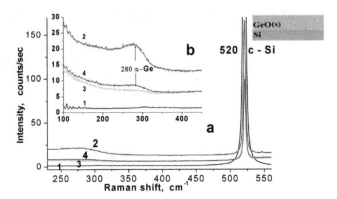

**Figure 14. a** –Raman spectra of GeO(solid) films of different thicknesses after decomposition of GeO(solid) into $GeO_2$ and Ge nanoclusters and oxidation of the nanoclusters: 1 – Si substrate, 2 and 3 - thick and thin areas of the GeO(solid) film, 4 - thick area of the GeO(solid) film where Ge nanoclusters were oxidized by the Ar + laser; **b** – magnified fragments of the same spectra.

Thus, the experiments with GeO(solid) and $GeO_x$ films showed that the modification processes of their structures proceeding during traditional anneals are qualitatively similar to PLA-induced processes. All such processes (film shrinkage and densification, all decomposition stages of the metastable GeO(solid) atomic network, formation of a-Ge nanoparticles and their oxidation and crystallisation) do not make the layers more transparent for Ar⁺ laser excitation radiation. Among all the above-considered physicochemical processes in the analyzed one- and bilayer compositions, only reactions of the glassy $GeO_2$ matrix in $GeO_2$<Ge-NCs> heterolayers with $Si_3N_4$ cap layers could lead to such an effect. Besides, the decrease of the Raman-spectrometer beam absorption in the Si/$GeO_2$<Ge-NCs>/$SiN_xO_y$ system with increasing the number of ns KrF excimer laser pulses given to the sample (Fig. 12) could be explained assuming that a-Ge nanoclusters formed by short laser pulses in the $GeO_2$<Ge-NCs> heterolayer were substantially smaller than a-Ge nanoclusters formed during traditional thermal treatments. Thus, the whole mass of free Ge atoms leaving the decaying atomic lattice of GeO(solid) during irradiations formed a multitude of very small c-Ge nanoparticles.

The latter is quite natural a situation under conditions of abrupt heating of the heterolayer not only with laser impulses, but also with the energy released during decay of the unstable atomic network of GeO(solid) undergoing decomposition during the discussed treeatments. However, due to the short width of laser pulses, the length of possible diffusion of free Ge toward Ge nanoclusters does not exceed 2-3 cluster diameters, as nanoclusters are separated with distances close to the mean radius of nanoclusters in studied heterolayers. Hence, in small-sized a-Ge nanoparticles, due to the 3D quantum confinement effect, the effective optical gap will be substantially wider than that of larger nanocrystals formed during traditional anneals and, hence, their radiation absorption effect will not be so high as that of the larger Ge nanoparticles.

The sensitivity of our Raman spectrometer has allowed us to reveal in our experiments specific features directly defined by the sizes of Si and Ge nanocrystals in thin-film materials (Volodin et al., 2005). In large Ge nanocrystals, the spectral position of the Raman peak is at 301,5 cm$^{-1}$, with its blue shift becoming notable at nanocrystal sizes < 6-7 nm. Data on the spectral position of the Raman peaks due to Ge nanocrystals hosted in a GeO$_2$ matrix as calculated from two theoretical models (data of (Nelin & Nilsson, 1972) and data calculated from the model of effective density of folded vibration states (Volodin et al., 2005)) are shown in Fig. 16. Raman spectra of irradiated GeO$_2$<Ge-NCs> heterostructures measured in the spectral region around the wavenumber 301,5 cm$^{-1}$ are shown in Fig. 12 b. In these spectra, a red shift of Raman peaks, typical of Ge NCs, is clearly observed. According to model calculations, the sizes of Ge NCs increase from 1,4 – 1,8 nm to 1,8 – 2,3 nm (Fig. 16)

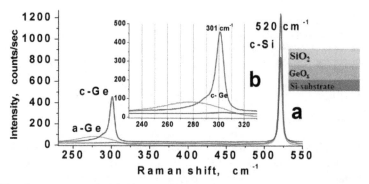

**Figure 15. a** - Raman spectra of a SiO$_2$/GeO$_x$/Si (100/100 nm) structure illustrating the precipitation of germanium in GeO$_x$ layers during anneals, and also the formation of a-Ge nanoparticles and their subsequent crystallisation: **1** - as deposited film; **2** and **3** – films annealed at 300 ºC for 40 min and at 480 ºC, 3 min + 530 ºC, 1 min, respectively; **b** – the same spectra at a larger magnification.

with increasing the number of ns KrF excimer laser pulses given to local regions on the sample surface. This result confirms the previous assumption on the reason for transparency enhancement for Ar+ laser radiation in GeO$_2$<Ge-NCs> heterolayers irradiated with 25-ns KrF excimer laser pulses ($\lambda$=248 nm).

Note that analogous treatments of non-stochiometric silicon-rich SiO$_x$ and Si$_x$N$_y$(H) glassy films with high-energy KrF excimer laser pulses ($\lambda$=248 nm, $t_p$=25 ns) activate a complex structure rearrangement process which begins with the emission of excessive Si atoms from the atomic network of the material due to breakage of Si-O bonds (in SiOx films) or Si-H and Si-N bonds (in Si$_x$N$_y$(H) films). The process ends with the formation of many a-Si nanoparticles in the film bulk from released Si atoms followed with subsequent crystallization of the nanoparticles (Gallas et al., 2002). In such cases, the silica films undergo shrinkage. Here, all processes proceed like in GeO$_2$<Ge-NCs> heterolayers. However, unlike in the latter case, in PLA-treated GeO$_2$<Ge-NCs> heterolayers there is no growth of transparency for Ar$^+$ laser radiation; on the contrary, according to (Korchagina et al., 2012), the absorption of 514,5-nm radiation increases. Today, the reason for the different behaviors demonstrated by the GeO$_2$<Ge-NCs> and SiO$_2$<Si-NCs> systems under PLA treatments remains unclear. In this effect, which depends on many factors, the contribution due to each factor changes under PLA differently in Ge- and Si-based heterosystems.

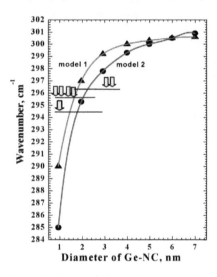

**Figure 16.** The spectral position of the Raman peak due to Ge nanocrystals versus nanocrystal size in GeO$_2$<Ge-NCs> heterolayers: ● – data of (Nelin & Nilsson, 1972), ▲ - data calculated by the model of effective density of folded vibration states (Volodin et al., 2005). Using the Raman spectra in Fig. 12, one can determine the sizes of Ge NCs size formed in local areas of the GeO$_2$<Ge-NCs>/ SiN$_x$O$_y$ bilayer system treated with one, two and four ns laser pulses.

Finally, we would like to emphasize here that irradiation of samples with 25-ns KrF excimer laser pulses ($\lambda$=248 nm) produces a very important effect on GeO$_2$<Ge-NCs> heterolayers, namely, its allows one to decrease both the sizes and size dispersion of Ge nanoparticles formed in metastable GeO(solid) and GeO$_x$ layers. This process proceeds in the strongly densifying GeO$_2$<Ge-NCs> heterolayer that consists of 70% vol. of GeO$_2$ matrix, which, during the same irradiation treatment, is capable of controllably changing its chemical composition, structure, and physical, optical and electrophysical properties.

## 4.2. Femtosecond treatments of samples with Ti-Sapphire laser

Differences in laser radiation parameters, and also differences in other irradiation conditions, strongly affect the material modification process and its ultimate results. The idea of to which extent the final effect due to PLA might differ in various irradiated systems can be grasped through making a comparison of PLA data for film structures involving GeO, GeO$_x$ layers, and GeO$_2$<Ge-NCs> heterolayers obtained in treatments performed using two different lasers, a KrF excimer laser and a Ti-Sapphire laser.

Phase transition mechanisms in solid materials under femtosecond laser treatments are radically different from analogous PLAs implemented using nanosecond pulses. These differences are most clearly manifested in the case of ultra-high-power femtosecond laser treatments, when nonlinear effects play an important role in the light absorption processes. There are many contributions devoted to the study of the interaction of femtosecond laser pulses with matter (Juodkazis et al., 2009). Results of an experimental study of the dynamics of electron-hole plasma formation and recombination of photo-induced charge carriers in Si under fs PLA at laser energy lower the threshold value for surface melting and ablation were reported by Ashitkov et al. (Ashitkov et al., 2004). During the action of a fs laser pulse, electron-hole plasma with electron concentration in the conduction band up to $10^{22}$ cm$^{-3}$ forms, both thermally activated and athermal surface-layer melting processes being observed along with a transition of material to metallic state. In (Ashitkov et al., 2004), time-dependent optical reflection was measured for heating-pulse energies.

In phase transitions under such conditions, athermal effects may emerge. Indeed, the duration of fs laser pulses is much shorter then the duration of the electron- phonon interaction in semiconductor (approx. 1-2 ps). That is why in silicon, during laser pulse the "hot" electron-hole plasma does not excite vibrational modes and relaxes insignificantly. The temperatures of the electron and atomic sub-systems are much different. According to theoretical calculations, the material becomes unstable when the electron concentration excited from the valence to conduction band reaches 9-20% of the atom concentration in Si lattice (Bok, 1981). This metastable state may relax to a more stable crystalline phase without melting yet with release of latent crystallisation heat. The process is similar to "explosive" crystallisation. In several picoseconds after the pulse, the phonon and electron temperatures should equalize. Post-pulse processes, e.g. post-pulse atomic diffusion in heated films, start manifesting. Very probably, diffusion can be stimulated by laser radiation not only via film heating, but also due to the breaking of valence bonds that occurs when electrons undergo excitation to the conduction band. Under nanosecond laser treatments, the time of Si film cooling to a temperature below the melting point is known to amount to about 100 nanoseconds (Sameshima & Usui, 1991). For fs laser treatments, the cooling time can be expected to be the same. This time can turn out sufficient for diffusion of excess silicon and for formation of nanocrystals only in the case of melt.

Zabotnov et al. (Zabotnov et al., 2006) showed that femtosecond laser annealing of Si surfaces leads to their nanostructuring, two types of formed nanostructures being observed. In one case, these are ordered lattices with a period somewhat shorter than laser radiation wavelength. The formation of such structures can be explained by the interference between the fs radiation waves and the non-equilibrium plasma excited by radiation in the surface layer. Here, ablation of Si atoms is accompanied with their rapid oxidation in air, and also with the formation of surface Si nanocrystals sized 3-5 nm.

Martsinovsky et al. (Martsinovsky et al., 2009) showed that, under the action of femtosecond laser pulses, intense photo-excitation of semiconductor surface, which dramatically modifies its optical response and provides conditions for generation of surface electromagnetic waves of various types, becomes possible. Martsinovsky et al. have also discussed the interrelation of electronic processes optically induced in the near-surface layer with the formation of periodical surface microstructures that were observed in experiments on irradiation of silicon targets.

Femtosecond PLAs were also used to crystallise silicon films. In PLA treatments, a critical point for thin films is proper choice of irradiation conditions since at high pulse energies such films may readily suffer ablation. At present, the issue of Si-nanocrystal formation in silicon nitride and silicon oxide films under nanosecond PLAs is a well-studied matter. A process for fabricating nanocrystals in dielectric films with the the help of nanosecond laser pulses was patented as involved in the production technology of non-volatile flash memory devices. Femtosecond treatments were used for modification of Si nanoclusters in SiO₂ films (Korchagina et al., 2012). Yet, for femtosecond treatments structural changes in silicon clusters in SiN$_x$ and SiO$_x$ films have not been studied. As for germanium, Ge films, and Ge nanoparticles embedded in dielectric layers, to the best of our knowledge, only one reported communication on fs-laser-initiated crystallisation in thin layers of amorphous Ge is presently available (Salihoglu et al., 2011).

On the whole, the physics underlying processes in film systems subjected to to femtosecond anneals still remains scantily understood. That is why it would be interesting to perform a study of such processes in samples treated with nanosecond and femtosecond laser pulses with revealing all involved characteristic features, also including possible athermal effect on Ge nanoparticles hosted in an insulator matrix, both with the aim of detecting new manifestations of PLA-induced effects in germanium-based materials and with the aim of comparison of similar processes in systems based on two chemically akin semiconductors, Si and Ge. Assumes that the results of these studies will prove also useful for applications.

### 4.2.1. Femtosecond PLA of GeO₂<Ge-NCs> heterolayers

Our study of fs PLA-treated thin-film systems based on germanium oxide layers was the first one devoted to the subject matter of interest. Therefore, the first task in such investigations was to develop a general concept of the complex effects and phenomena occurring in the film systems of interest subjected to PLAs under different conditions. In particular, we examined GeO₂<Ge-NCs> heterofilms of different thicknesses grown by the LP CVD method on (bare and CVD SiO₂-covered) silicon or quartz glass substrates. The films either had one layer and no cap layer or they were protected by a thin cap layer of PE CVD SiN$_x$O$_y$ or CVD SiO₂. In another set of samples, ~ 20 – 30 nm and ~200-250 nm thick, GeO₂<Ge-NCs> layers were part of multilayer compositions where they were separated with thin (~7-10 nm) and thick (~100 nm) CVD SiO₂ layers as shown in Fig. 17 a. Scanning was made with a focused fs-laser beam at a velocity of 100 μm/s. The frequency of laser pulses was 10 pulses per 1 μm. The laser fluence ($E_{pulse}$) varied from 15 to 1 mJ in ~70-μm diameter spot and from 0.028 to 0.0018 mJ in 1-μm diameter spot. In the experiments, samples with GeO$_2$<Ge-NCs> heterolayers were used. In addition, in some cases optical bench vibrations of ~1,5-μm amplitude in two dimensions, in the plane of the optical bench and in the direction normal to it, were implemented. A typical result for layered systems with GeO$_2$<Ge-NCs> heterolayers on Si substrate scanned with Ti-Sapphire laser beam focused to a 70-μm diameter spot is shown in Fig. 17 b. From line to line, the laser fluence was decreased in 1-mJ steps within 15-mJ wide intervals of $E_{pulse}$. Thus, the adopted femtosecond PLA regimes and conditions for examined films were different from their treatments with the nanosecond laser in terms of the majority of basic physical parameters (laser radiation wavelength, pulse duration, intervals between pulses, arrangement of the irradiation process). Besides, in the case of fs PLAs the adopted irradiation technique did not allow us to trace changes in the film structure following successive irradiations of the sample with a series of laser pulses.

The first result was that optical microscopy was quite a sensitive technique allowing us to reveal both changes in the surface relief and the variations in the optical parameters of the material induced by structural and chemical modification processes proceeding in heterolayers under laser beam treatments, whereas the SEM method was efficient only when studying surface morphology modifications (cp. Figs. 17 a and b). The main fundamental difference in the PLA mechanisms for the two lasers is manifested when the

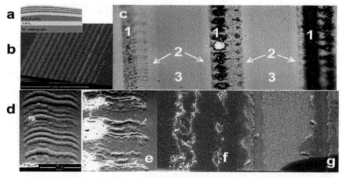

**Figure 17.** Surfaces of multilayer coatings involving alternating GeO$_2$<Ge-NCs> heterolayers and SiO$_2$ layers after fs PLA (beam diameter 70 μm): **a** – schematic of the multilayer coating (similar to that in Fig. 6); **b** – SEM image of the film coating treated with laser pulses of various energies; **c** – optical microscopy image of the same sample: (1) evolution of surface morphology at the centerline of laser spot path; (2) along coating edges there is no relief, although light stripes defined by the changes of optical properties of some layers inside the coating are clearly seen; (3) laser treated paths are separated with wide stripes where the film coating structure has remained unchanged ($E_{pulse}$ was not high enough); **d, e, f** and **g** – SEM images obtained at the various stages of the coating laser damage process implemented with laser energy increase: **d** – development of a wavy surface relief as a result of transformation of 3 thin GeO$_2$<Ge-NCs> heterolayers in the top tier into nanofoam ($E_{pulse}$ = 8 μJ ); **e** – beginning of layer destruction in the top tier ($E_{pulse}$ = 9 μJ ); **f** – simultaneous damage of upper and middle tiers ($E_{pulse}$ = 10 μJ); **g** – simultaneous damage of all layers down to the Si substrate and the beginning of its damage by formed ripples ($E_{pulse}$ =13 μJ).

laser are used to modify GeO$_2$<Ge-NCs> heterolayers. Only Ge-nanoclusters, whose optical gap $E_g^{eff}$(Ge-NC) < 2 eV, absorbed femtosecond laser radiation with $\lambda$=800 μs in GeO$_2$<Ge-NCs> heterolayers (see Figs. 8 and 9). Nanosecond KrF laser radiation with $\lambda$=248 nm is additionally absorbed by the GeO$_2$ matrix, which constitutes 70% of the heterolayer volume, its optical gap being $E_g^{eff}$(GeO$_2$) ~4,5-5 eV. The actions of the light emitted by the two lasers on the two material components in the heterolayers are radically different: while KrF laser radiation heats both the Ge-nanoclusters and the dielectric matrix, Ti-Sapphire lase radiation heats only Ge-nanoparticles while leaving the matrix cold. It is this difference that makes the effects due to PLA processes in GeO$_2$<Ge-NCs> heterolayers under irradiation with the two lasers cardinally opposite. Namely, in GeO$_2$<Ge-NCs> heterolayers treated with the fs laser the heated Ge-nanoclusters begin to react with the surrounding glassy GeO$_2$ and, during this process, the initial heterostructure transforms into a totally different kind of solid with fundamentally different physical properties.

### 4.2.2. Formation of GeO$_2$ nanofoam using fs PLA of GeO$_2$<Ge-NCs> heterolayers

Indeed, Ge-nanoclusters surrounded by GeO$_2$ matrix were rapidly heated by fs laser radiation, to which the oxide was transparent, unless the chemical reaction (4) between both components in GeO$_2$<Ge-NCs> heterostructure was initiated. The emitted germanium monooxide molecules (GeO(gas)) are initially localised in the vicinity of hot Ge-nanoparticles surrounded by a cooler insulator (Fig. 18). Affected by the vapor temperature and pressure, the glassy GeO$_2$ matrix gets heated and it extends when softening. On cooling, the GeO$_2$<Ge-NCs> heterostructure turns into a swelled mass of nano-dimensional glass cells. Each of the initial Ge-nanoclusters, as its grows smaller and lighter, is to remain inside a formed nano-cells (Fig. 18 c). The most surprising thing here is that, in the outcome of nano-foam formation process, all Ge-nanoparticles should have a uniform minimal size independently of their size in the initial heterolayer. Due to the 3D quantum-dimensional effect, Ge-nanoclusters, as they grew smaller during their reaction with the GeO$_2$ matrix, should stop absorbing the heating light according to the condition $h\nu = E_g^{eff}$(NCs). This technique allows us to obtain ensembles of isolated indirect gap semiconductors of very small dimensions that also have a low size dispersion.

The process is completed as follows. While the nano-cell sizes in the glass softened under the internal gas pressure of the gas inside grow in value, their glass walls start cooling. The latter is due to the fact that, first, the gas in the nano-cells volumetrically expands under near-adiabatic conditions and, hence, the gas temperature will decrease. In addition, the mechanical expansion of the nano-cells in the soft GeO$_2$ nanofoam consumes the energy from the gas in cells. The gas gives off heat and its pressure decreases. Simultaneously, the nanocell-hosted Ge-nanoclusters, as they grow smaller, start absorbing less fs laser radiation (due to the 3D quantum-confinement effect). This reduces the efficiency of the reaction yielding new portions of GeO(gas) in the nano-cells. Thus, in the nanofoam the nano-cell walls start cooling at a sufficiently high internal pressure due to GeO(gas) vapor. Cooling of the GeO$_2$ glass rapidly increases its viscosity, which prevents the nano-cells from being compressed with decrease of the internal GeO(gas) pressure. Thus, the individual nano-cells, as well as all of the nanofoam, will preserve their dimensions in the cooling process.

PLA results of multilayer film systems consisting of a conventional dielectric layers and $GeO_2$<Ge-NCs> heterolayers illustrate in Fig.17 the complex effect of fs laser radiation on the system. The system consisted of nine layers belonging to two types: thick and thin $SiO_2$ films and $GeO_2$<Ge-NCs> heterolayers that alternated among each other (according to the diagram in Fig. 17 a). With increase of laser pulse energy, distinct indications of film structure modification started showing up, becoming manifested more and more clearly. The sample holder vibration accompanying the scanning process proved useful in revealing important details of the mechanisms underlying the impact of laser pulse radiation on the structure of irradiated films. Changes in film properties began to evolve in the volume of the films, and then they continued in surface morphological changes. To describe the observed processes, it was necessary to employ the whole complex of available experimental methods, each method, as a rule, allowing characterisation of a limited number (one or two) of the changing properties of the objects under study.

For instance, optical microscopy (Fig. 17 c) was sensitive to variations of optical parameters of the heterolayers under laser beam, manifested in color changes. This was an indication of the beginning of modification of material properties in some layers inside the coating. Those processes preceded the changes of the thickness and surface relief of the coating. SEM was effective only when studying surface morphology modifications (cp. Figs. 17 b, d, e, f and g). Besides, it has limited capabilities in height measurements of many relief details. However, when combined, the two methods were capable of revealing a complete picture of the dynamics of fs PLA induced changes in investigated multilayer structures on increasing the laser pulse energy.

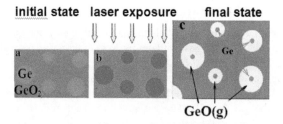

**Figure 18.** Schematic illustrating the process of transformation of a $GeO_2$<Ge-NCs> heterolayer in nanofoam-like glassy $GeO_2$ in a film structure subjected to fs laser annealing: **a** - initial heterolayer structure; **b** - heating of Ge-nanoclusters by fs laser pulses, which initiates the formation of GeO(gas) around these clusters; **c** - final structure of nanofoam-like glassy $GeO_2$.

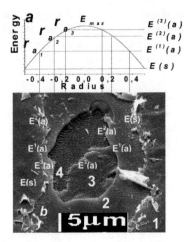

**Figure 19.** The simultaneous laser damage of two tiers of the multilayer system in Fig. 17 g during fs PLA (SEM image): **a** - radial Gaussian distribution of energy inside the laser beam; **b** - $E^{(1)}(a)$, $E^{(2)}(a)$, $E^{(3)}(a)$ - threshold energies for laser damage: (1) - in the top tier of the multilayer film system, (2) and (3) - in a thick SiO₂ layer and in a thick GeO₂<Ge-NCs> heterolayer of the middle top, respectively; $2r_{a1}$, $2r_{a2}$, $2r_{a3}$ - width of damaged areas in the respective layers of the system; $E(s)$ - threshold energy for the onset of the foaming process in thin GeO₂<Ge-NCs> heterolayers of the top tier of the multi-layer film system.

The experiment revealed some interesting features of the nanofoam formation process. First of all, the process was indeed implemented. The latter was indirectly proved by several methods of structural analysis. For instance, at an early stage of the process a wavy relief not violating the continuity of the multilayer coating emerged on the coating surface. The formation of this relief was registered by optical microscopy and SEM (Figs. 17 c and d). With increasing the laser pulse energy, the top tier in the coating starts destructing, and the smooth thick SiO₂ layer of the middle tier stands out. Then, the top tier of the coating detaches from the lower tier, tears and rolles, with all its thin layers becoming laminated (see Figs. 17 f and 19 b). This is the first stage of the laser damage of the whole multilayer system. Further on, the process laser damage develops involving more and more layers (Fig. 17 f and g and Fig 19 b), the boundaries of destructed zones in each layer spreading from the centerline of the laser treated path on the sample surface with growth of laser pulse energy. The ablated zone (band) in each layer is always wider than that in the lower layer. Therefore, SEM images allow observation of the structure and morphology of the uncovered surface of any exposed coating layer down to the Si-substrate surface.

Ultra-short laser pulses of the investigated heterostructures cause structural rearrangements and chemical reactions as: films shrinkage, GeO(solid) layers decay on reaction (3), growth of Ge-nanoparticles in the heterolayer and their crystallisation, chemical reactions of GeO₂ matrix of heterolayers with cap layers on their interface, possible crystallisation of GeO₂ glass around Ge-nanoparticles and reaction of GeO vapour formation from GeO₂<Ge-NCs>

heterolayer components, also ablation — explosive damage of films and substrates (Korte et al., 1999). Many of these processes are also expressed in radiation of GeO$_2$<Ge-NCs> heterolayers by fs Ti-Sapphire laser, just as in case of Kr excimer laser. But, as is seen from Fig. 17, the interaction character of fs laser radiation with the studied films and substrate was considerably different.

Results of using PLA and fs lasers for the studied film coatings showed that a number of the above-mentioned structural rearrangement processes and chemical reactions excite one by one with laser fluence growth. In general, to activate a new (i)-process, it is necessary to have laser fluence, absorbed by a film coatings square unit, exceeding the energy threshold of (E$^{(i)}$). Energy intervals among thresholds of different processes successively excited in films are usually so large that they can be definitely separated from each other. Therefore, the processes observed in multi-layer systems could be divided into two main groups. The first one consists of almost all the processes from the presented list, which proceed in the volume and interfaces of each of a multi-system's layers independently from processes in its other layers. Ablation processes belong to the second group as they, proceeding in one of a multi-system's layers, as a rule, determine all other processes in its lower layers.

### 4.2.3. Laser damage in multilayer coatings

In investigated heterostructures, ultra-short laser pulses induce various structural rearrangements and chemical reactions such as film shrinkage, decomposition of GeO(solid) layers according to reaction (3), growth of Ge-nanoparticles followed with their crystallisation in the heterolayer, chemical reactions of the GeO$_2$ matrix with the cap layers in the interfacial region, possible crystallisation of GeO$_2$ glass around Ge-nanoparticles and reaction (4) of GeO vapor formation from GeO$_2$<Ge-NCs> heterolayer components, and also ablation and explosive laser damage of the films and substrates (Korte et al., 1999). Many of these processes are manifested in GeO$_2$<Ge-NCs> heterolayers irradiated with fs Ti-Sapphire laser in a similar way like in the case of Kr excimer laser. However, as it is seen from Fig. 17, the interaction of fs laser radiation with examined films and Si substrate was largely different.

Results of using ns and fs PLA for treating examined film coatings showed that a sequence of the above-mentioned structural rearrangement processes and chemical reactions gets initiated on increasing the laser fluence. In general, for activation of a new (i-th) process, it is required to reach a laser fluence absorbed by a film coatings in unit square in excess of the energy threshold for the process (E$^{(i)}$). The energy intervals between the thresholds of the various processes successively initiated in the films are usually so large that they can be definitely separated from each other. Therefore, the processes observed in our multilayer systems could be divided into two main groups. The first group consists of almost all the processes from the above process list, which proceed in the bulk and interfacial regions of each of the involved layers independently of the processes proceeding in other layers. Ablation and laser damage processes belong to the second group as they, proceeding in one of the layers, as a rule, govern the processes proceeding in lower layers.

Successive onsets of structural-chemical modification processes of group I was observed in bilayer coatings given ns PLAs. In the case of fs PLA, successive (layer-by-layer) initiation of structural-chemical modification processes becomes distinctly pronounced during ablative destruction of multilayer coatings (see Fig. 19 b that shows a fragment of Fig. 17 f). It is seen that the laser pulse fluence absorbed by the multilayer coating is sufficiently high to exceed four energy threshold values, $E(s)$, $E^{(1)}(a)$, $E^{(2)}(a)$, and $E^{(3)}(a)$ (see Fig. 19 a). Three of the threshold energies are the thresholds for laser damage initiation in the two top tiers of the irradiated system (in the top tier and in the two layers of the middle tier, a thick SiO₂ layer and a GeO₂<Ge-NCs> heterolayer). The very first threshold energy is the non-ablative threshold $E(s)$ of the foaming process of three thin GeO₂<Ge-NCs> heterolayers in the top coating tier. The radial distribution of light energy in the laser beam obeys the Gauss law $E(r) = E_0 \exp(-2r^2 / r_0^2)$ (Korte et al., 1999) as shown in Fig. 19 a; therefore, destruction of the layered structures begins at the centerline of the laser treated path on the film surface when the pulse energy exceeds the necessary energy threshold of the top tier, $E^{(1)}(a)$ (see Fig. 19 b). If the pulse energy is sufficient to exceed the damage thresholds of the first, second and subsequent film, the laser damage will be initiated in all these layers simultaneously. Yet, ablative removal of the lower- coating material is impossible without a preliminary removal of the upper films. That is why, in most cases multilayer coatings are damaged successively, layer by layer. In this case the positions of the boundaries in destructed areas of each of the layers are determined by the radial energy distribution in the laser-beam cross section. Qualitatively, the theoretical expectation for this regularity is illustrated by Fig. 19 a. The experimental data of Fig. 17 f, g, and Fig. 19 b, in principle, do not disagree with the stated concept, although a multi-mode laser beam structure in real cases often distorts the strictly circular pattern of radial energy distribution in laser spots on irradiated surfaces. Thus, the processes activated in PLA-treated films experimentally revealed the real energy distribution round the laser-beam center.

The particular case under consideration is interesting in that respect that, apart from the propagation of the laser damage process into depth in a multilayer coating involving two types of films of different origin, here there was also a process that can be classed to intra-layer processes that can also proceed in layers covered by other films. This was the transformation process of fs PLA treated GeO₂<Ge-NCs> heterolayers into a nanofoam. The formation of nanofoam in germanate heterolayers was not hindered by thin SiO₂ layers that covered such heterolayers. Such SiO₂ layers were weakly heated by the laser beam, and they do not change their properties due to chemical and structural modifications. Simultaneously, this process progressed along the centerline of the fs laser treated path when the energy threshold of nanofoam formation E(s) was overcome during the fs PLA. If the laser fluence many times exceeded the threshold energy $E(s)$, then the top tier was immediately torn off from the middle tier as a homogeneous film according to the common explosive machanism of film damage. Here, the chemical nanofoam formation process did not have sufficient time to evolve in the GeO₂<Ge-NCs> heterolayers of this tier. Here, under fs laser irradiation the system of upper layers behaves as a solid film of homogeneous material. The thickness of this layer system at its edges along the line of damage rupture was almost no difference from the thickness of the whole layer on the substrate, and the edges of the layer were like those of glassy films under brittle fracture. In other words, the edges mostly looked like broken line segments (Fig. 20 b, area 1), which were similar to the edges of thick SiO₂ layers after damage ruptures in the same multilayer coating (Fig. 20 b, areas 2 and 4).

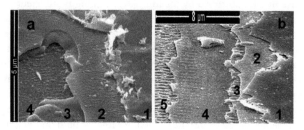

**Figure 20.** SEM image of layer-by-layer destruction processes in a multilayer structure exposed to laser radiation ($a - E_{pulse}$ = 10 mJ , $b - E_{pulse}$ = 13 mJ) : **1** - the top tier of thin layers, **2, 3** – the middle tier of thick layers (SiO₂ and nanofoamed GeO₂<Ge-NCs> heterolayer, respectively), **4, 5** – the bottom tier (lower SiO₂ layer and Si substrate, respectively.

However, laser damage took a different course if, under fs PLA, the laser fluence was only a little in excess of the threshold E(s) and the process of GeO₂<Ge-NCs> heterolayer foaming had sufficient time to develop. The thicknesses of those layers increased appreciably together with the thickness of the whole top tier in the multilayer coating; the top tier swelled and began to gradually tear off from the underlying SiO₂ layer. Then the areas of the top tier showing a most pronounced nanofoam formation suffered destruction accompanied by tearing off of its fragments (see Figs. 19 b and 20 a). Here, there are many thickened monolayer fragments of the top tier, exfoliated from the substrate incompletely and left in many areas over the edges of the layer along the centerline of the damage rupture zone. Those fragments roll in microtubes and laminate into very thin separate layers that constitute this tier (Figs. 19 b and 20 a).

During fs PLA, the nanofoam formation process of the thick GeO₂<Ge-NCs> heterolayer in the middle tier of the film coating proceeded in a different way than that in thin GeO₂<Ge-NCs> heterolayers of the top tier. The reason for this was that the thickness of GeO₂<Ge-NCs> heterolayer and its mass allowed it to accumulate more laser energy per unit volume. Moreover, a longer time was required for the damage process to cause removal of the thin film layers and the upper thick SiO₂ layer. Both factors allow the heterolayer to accumulate in its volume a considerable energy and considerably prolong the action of this energy transforming the heterolayer into nanofoam. Therefore, in all SEM images (Figs. 17 g, 19 b, 20 a and b) one can observe under the upper thick SiO₂ layer removed by laser damage a substance that looks like a liquid light foamy mass.

### 4.2.4. Some characteristic parameters of the nanofoam formation process

The thicknesses of the layers in the bottom and middle tiers as measured in SEM images were close to the values obtained in as-synthesized films. We could also conveniently measure in the SEM images the thicknesses of the foam-like layers, which were found to 1,5-3 times exceed the initial thicknesses of the GeO$_2$<Ge-NCs> heterolayer. In some areal parts of the foam-like layers (where the layers became uncoated following the laser damage of the SiO$_2$ layer, Fig. 20), the foamy mass often increased, approximately by 2-4 times, the thickness of the GeO$_2$<Ge-NCs> heterolayer. During of the most foamed layer, some individual fragments of that layer were often explosively thrown up onto the SiO$_2$ layer. The thicknesses of those fragments were bigger than the three- or four-fold thickness of the SiO$_2$ layer.

As a rule, AFM data on the sizes of important relief features of the film coatings (Fig. 21) complied with the data obtained by other measuring techniques. Thus, AFM measurements proved that the top tier tears off from the underlying SiO$_2$ layer during laser damage ablation (Fig. 22). Besides, it is possible to determine the initial thickness ($h$) of the top tier ($h$ ~50-60 nm) and the degree of its changes during fs PLA using the surface relief profile (Fig. 22 b) measured along the $p$-line (Fig. 22 c). The obtained value for the thickness of the top tier ($H$) in the immediate vicinity of the damage rupture (Fig. 22 a, area 2) is five times greater than the initial layer thickness ($h$). But the value H is not only the result of transformation of the three GeO$_2$<Ge-NCs> heterolayers in the top tier into nanofoam. The laminating effect of the edges of the top-tier layers from the lower films in this area also contributed to the found value of H since after the tearing the free edges bent upwards.

The energy threshold of the foam formation process $E(s)$ in the thin GeO$_2$<Ge-NCs> heterolayers of the top tier during fs PLA was exceeded over the whole irradiated film surface (Figs. 21 and 22). Moreover, the rate of irradiation along the laser treated path (close to the line $q$ in Fig. 22 a) varied periodically and at the points of maximum it reached the lower threshold energy for laser damage in this tier $E^{(1)}(a)$. In accordance with these periods, crests in swollen and buckled regions of the top tier formed. On the average, the separation between crests is 2,4 - 2,5 $\mu$m (line $q$ in Fig. 22 a). There are zones in between the crests, where there was a minimal foaming of GeO$_2$<Ge-NCs> heterolayers. From the relief profile measured along the cutting line $q$ (Fig. 22 d), we could determine the modulation frequency

**Figure 21.** 3D AFM image of fs laser treated path on the multilayer surface shown in Fig. 17 e.

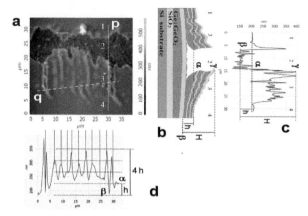

**Figure 22. a** - AFM image of the surface of the multilayer coating shown in Fig. 21: *p* and *q* - cutting lines of the sample surface across and along the fs laser treated path, respectively; **b** – schematic of the sample cross section along the line *p* drawn along the rupture edge of the the top tier of the coating with swelled films (**1** and **4** - areas with unchanged relief, **2** - laser damage zone of the top tier, **3** - zone of the maximum change in relief at the edge of rupturing film during its swelling; **c** - relief of the film coating along the line *p*: $\alpha$ - the initial level of coating surface, $\beta$ - surface of the upper thick $SiO_2$ layer, $\gamma$ - the highest level of coating relief changes, **d** - relief of the film coating along the line *q*.

of laser beam energy connected with the sample vibration. An analysis of this relief in the described fs PLA conditions showed that, at this modulation frequency of laser irradiation, its energy changes within the interval between the threshold energies of the two processes $E^{(1)}(a)$ and $E(s)$ during the period of ~ 0,025 sec. The thickness of $GeO_2$<Ge-NCs> heterolayers may increase 3-4 times for the time equal to half the period.

As a result, the vibration experiment revealed: (i) a low threshold energy $E(s)$ of the process; (ii) a high capability of $GeO_2$<Ge-NCs> heterolayers to increase of their volume (by 3-5 times) during transformation into nanofoam material; (iii) high dynamic characteristics of the nanofoam formation process during increase of fs laser pulse energy.

Micro-Raman spectra of fs laser treated $GeO_2$<Ge-NCs>heterolayers (Fig. 23) produce another evidence of their transformation into a nanofoam-like material. For simplification of the analysis, the spectra were registered in $GeO_2$<Ge-NCs> heterolayers deposited onto a Si substrate protected by a 25-nm thick $SiN_xO_y$ cap layer (the effect due to ns PLA was also studied on this sample). The effect due to laser fluence value exciting the nanofoam formation process in local regions of the $GeO_2$<Ge-NCs>/$SiN_xO_y$ bilayer system is shown in Fig. 23. In Fig. 23 a, in spectrum 1, which corresponds to non-irradiated film at point 1 (Fig. 23 e), only a broad Raman band due to light scattering by transversal optical phonons of $\alpha$-Ge is shown, like in the case of ns laser treatment of this film system (Fig. 12).

As it can be judged from the weak scattering peak of the beam of the exciting $Ar^+$ laser of Raman spectrometer due to the Si-substrate lattice (scattering band in the region of 520 $cm^{-1}$), the film system was little transparent to this beam. Raman spectra 2 and 3 in Fig.

23 a were obtained at two closely located points (2 and 3) on one of the lines treated by the fs laser beam (Fig. 23 d and e, respectively). Yet, because of heterogeneity of the radial energy distribution in the beam, intensive formation of nanofoam mass from the GeO₂<Ge-NCs> heterolayer proceeded at both points of this line during PLA. At point 3, the film coating was at an early stage of destruction, still remaining almost intact, whereas at point 2, damage was well activated. The latter is evident from the intensity of the Raman peaks of Ar⁺ laser beam due to Si-substrate lattice (in the corresponding spectra); at point 2 the film practically does not hinder the Ar⁺ laser beam to reach the substrate. In other words, at points 2 and 3 we observe an early and the final stage of the destruction process of the foamed layer. Nonetheless, despite the possibility of mixing up film fragments belonging to different areal parts of the film, Ge-nanocrystal sizes were not large, ~2,3 – 3,1 nm as calculated by model (1) and ~3,1 - 4,1 as calculated by model (2). At point 4 in Fig. 23 c the fs laser energy was slightly lower than that at points 2 and 3. Therefore, the film foamed without and violation of its continuity. Ge-NCs of the least diameter ~1,0 (model (1)) and 1,4 nm (model (2)) were found here (spectrum 4 in Fig. 23 a).

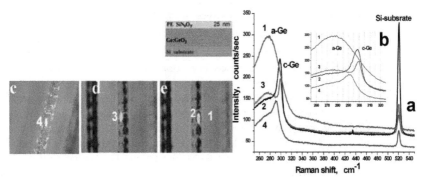

**Figure 23. a, b** - Raman spectra and **c, d, e** - photos of GeO₂<Ge-NCs> heterolayers transformed to nanofoam under a SiNₓOᵧ cap layer on the Si substrate: **1** – non-irradiated films; **2** - strong laser damage of the films; **3** - laser damage at an early stage; **4** - the formation of foam without destruction of the cap layer.

Thus, the both main features which are to accompany the formation of nanofoam from GeO₂<Ge-NCs> heterolayers during their fs laser pulse treatments were observed in the experiment. Direct images of the studied material are supposed to be obtained with HR TEM, but additional information about the properties of the nanofoam was also obtained with the SEM method. Characterizing the most impressive effects produced in GeO(solid) films and GeO₂<Ge-NCs> heterolayers by fs PLA, we would like to note that, here, the same structural and chemical modification processes like those in ns PLA treated samples proceed both consequently and in parallel in the material; those process include: decomposition of the metastable solid germanium monoxide atomic network with its transformation into the atomic network of glassy GeO₂ oversaturated with Ge atoms; clusterization of released germanium into amorphous nanoparticles; and growth and crystallisation of α-Ge nanoparticles. The effect of film transparency increase is also common for both kinds of laser treatments during crystallisation of $\alpha$-Ge nanoparticles in the film bulk. On the other hand, the main difference between the effects due to UV ns and IR fs PLAs on GeO$_2$<Ge-NCs> heterolayers consists in that UV ns laser radiation leads to a considerable shrinkage of the material, whereas IR fs radiation leads to transformation of the whole irradiated heterolayer material into a very loose foam containing nanocavities separated with ultra-thin glassy GeO$_2$ barriers. An SEM image of the nanofoam obtained by fs PLA of GeO$_2$<Ge-NCs> heterolayers unprotected by cap layers (Fig. 24 d) confirms such a microstructure of the obtained material.

Another remarkable feature is that the ablation and laser damage of the multilayer system proceeding together with the formation of nanofoam in GeO$_2$<Ge-NCs> heterolayers (Fig. 17 and 23) are processes highly sensitive to even small variations in the fs laser fluence distributed over the surface of the multilayer. On the contrary, the shrinkage process of the same multilayer system (illustrated by Figs. 7 c and 10) under ns pulsed laser radiation, even at the highest energies, demonstrates high uniformity of the irradiated coating surface. In GeO$_2$<Ge-NCs>/Si$_x$N$_y$O bilayer films consisting predominantly of GeO$_2$<Ge-NCs>, no ablation or laser damage effects were observed in the examined ranges of heterolayer thicknesses and laser fluences.

The causes for this difference may be somehow related with the fact that the shrinkage of heterolayer thickness is a process that proceeds at a rate ten times slower in comparison with the nanofoam formation process. The volume of a foaming heterolayer normally increases by a few hundred per cent, whereas the maximum shrinkage of film thickness (in the case of ns laser treatment) does not exceed 40 per cent. It is possible that during the latter relatively slow process local temperature non-uniformities get smoothed over the film. In other words, temperature and stress gradients have enough time to disappear from the film. Another cause may be related to the fact that the mechanisms of laser radiation absorption by the material in the two cases are very different. In GeO$_2$<Ge-NCs> heterolayers, the ultraviolet radiation emitted by ns pulsed laser is readily absorbed not only in Ge NCs, but also in the glassy GeO$_2$ matrix. As a result, the matrix, absorbing the main part of the incident light, undergoes quick heating. Ge NCs absorb less energy of light pulses in the heterolayer as they occupy only about 30% of its volume. Microscopic regions that do not absorb laser radiation are absent from heterolayer, making the distribution of absorbed heat in the film volume less inhomogeneous. If the average temperature in heterolayer does not exceed the energy threshold for GeO(gas) formation due to reaction (4) at the interface between the GeO$_2$ matrix and Ge-nanoclusters, the heterolayer shrinkage process will mainly depend on the matrix viscosity. The temperature of the film defines the film material viscosity. Then, it can be anticipated that the shrinkage process in the film will most likely proceed similarly to the shrinkage during traditional anneals of ordinary CVD SiO$_2$ and SiO$_3$N$_4$ films (as described in Section 3.3). Due to this shrinkage mechanism, much of the internal lateral stresses turn out to be reduced by viscous flow of the material. The lower the viscosity the more readily the film shrinking process proceeds.

In the case of fs laser treatments of GeO$_2$<Ge-NCs> heterolayer, radiation will be absorbed only in Ge-nanoclusters. As a result, the distribution of the heat absorbed in the heterolayer should be much less homogeneous in comparison with the case of ns laser pulses. Temperature gradients in heterolayer, which are created by fs laser pulses between the cold glass and the heated Ge-nanoclusters can be expected quite high because all the energy of the light absorbed in the heterolayer is transferred into the heterolayer bulk only through the boundaries of Ge-NCs with the glass matrix. Thus, in this system only a small part of the glass directly bordering on heated Ge-nanoclusters can be quickly heated to high temperatures, while the rest part of the glass matrix will undergo heating at a much slower rate. Such a state of the heterogeneous material will be characterised by abrupt spatial variations of temperature and compressive stress over the volume of the heterolayer occurring over distances comparable with Ge-NC sizes. In particular, effects associated with thermal expansion coefficients of GeO₂ glass and Ge-NCs, as well as with the formation of high-pressure GeO(gas) bubbles around Ge-clusters, will facilitate smoothening of internal stress inhomogeneity. The impact action of fs laser pulses stimulates rapid evolution of the latter state of the material in heterolayer.

## 4.3. Prospects for use of GeO₂<Ge-NCs> heterolayers in medicine and laser lithography

High sensitivity to laser irradiation readily activating the evaporation reaction of GeO(solid) films and GeO₂<Ge-NCs> heterolayers proceeding with the formation of volatile germanium monoxide is an interesting prerequisite for using such films as a nanoresist in laser lithography. Therefore, some experiments were carried out to evaluate the potential offered by the examined films in this field. In our study, GeO(solid) films and GeO₂<Ge-NCs> heterolayers not covered with cap-layers were used. Such films and heterolayers were deposited onto pure Si substrates or onto SiO₂ layers grown by CVD on Si. Then, the films were subjected to fs Ti-Sapphire laser PLA in the same way like in previously described experiments with multilayer coatings, except for the laser beam was focused in ~1-μm diameter spot and the pulse energy falling onto the sample was decreased with the help of optical filters. The maximum laser-beam energy ($E_{max}$) was chosen to exceed the threshold energy ($E^{(1)}(nfoam)$) of the foam formation process in the heterolayer yet it was lower than the threshold energy $E^{(2)}(evap)$ of the evaporation reaction of the film GeO (Fig. 24 a). Such a beam was used to scan the surface of a heterolayer on a Si substrate covered with a SiO₂ film. The sample was mounted on an optical bench vibrating in the lateral plane and vertically within 1 - 1,5 μm. In the sample vertical vibration, the laser beam focus was on the film at one turning point 1 of the oscillation period, and the focus raised over the film surface during the reverse motion of the optical bench. In the first case, the beam had a minimal diameter on the sample ($2r_{o1} \sim \lambda$) and a maximal energy value ($E'_{max}$) which was higher than the evaporation threshold of the film $E^{(2)}(evap)$ and its foaming threshold $E^{(1)}(nfoam)$ (i.e. $E'_{max} > E^{(2)}(evap) > E^{(1)}(nfoam)$). The defocused beam in point 2 had the biggest diameter $2r_{o2} > 2r_{o1}$ and a minimal value of the energy maximum (i.e. $E''_{max} < E^{(1)}(nfoam)$ and $E^{(2)}(evap)$).

The results of such scanning of the used GeO₂<Ge-NCs> heterolayer ~ 60-80 nm thick with a vibrating beam spot are shown in Fig. 24 b and c. The track width periodically varies due to pulse energy modulation by the sample vertical vibration. No changes in film material properties were observed within the laser treated path at point 2 (Fig. 24 b) as the laser pulse energy on the layer surface was minimal here. The foam formation process is initiated in the heterolayer when the laser pulse energy on its surface grows in magnitude along the laser treated path due to light beam narrowing. The reaction zone eventually broadens up, and the film thickness at the beam centre increases by ~2 - 3 times. At the moment when the rate of the process is maximal, the heterolayer evaporation energy becomes exceeded (point 1 in Fig. 24 b) as the energy $E'_{max}$ in the experiment was a little higher than $E^{(2)}(evap)$. Evaporation begins at the beam centre, analogously to the reaction of foam formation. The stretch of laser treated path over which this reaction proceeds is not long as the distance between the focusing lens and the sample changed due to vibrations simultaneously with the lateral shift of the beam over its trajectory. So, the beam on the sample surface again becomes defocused, and the nanofoam evaporation process ceases. In this case, the width of the cavities (~11-50 nm) formed by the beam in the foamed layer is quite appreciable (Fig. 24 b and c). This result attracts attention, as in geometric optics a spot having a diameter less than ~$\lambda/2$ is considered to be the diffraction limit of beam focusing, and the traces of its effect for photoresist should be close to this size considered as the physical limit for laser lithography line width. Through-holes shaped as round dots or prolonged windows, whose transversal sizes were ten times larger than the expected limit were obtained in our experiments with laser radiation wavelength $\lambda$=800 nm.

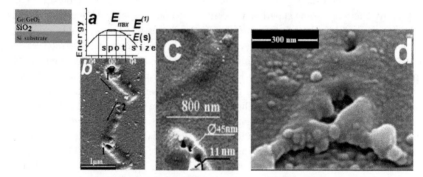

**Figure 24.** SEM images of the local formation of nanofoam in a GeO$_2$<Ge-NCs> heterolayer without a cap layer after fs laser irradiation (beam diameter ~ 1 μm;): **a** - radial Gaussian distribution of light energy in the laser beam; $E(s)$ and $E^{(1)}(evap)$ are the threshold energies for foaming and evaporation of the heterolayer material, respectively; **b** and **c** - focused beam treated paths on the heterolayer surface during horizontal and vertical microvibrations of the optical bench; **d** – nanofoam.

The mechanism behind the laser beam effect on the GeO$_2$<Ge-NCs> heterolayer has not been adequately understood. Possible explanations of the sharp narrowing of the high-flux zone in the laser beam section on the film surface were considered, including intermode interference of light rays inside the laser beam and the effect of two-photon absorption of light by the film in areas with high degree of radiation power localization on the film surface.

Note that in this experiment no additional measures were taken to stabilize the irradiation conditions of GeO(solid) films and GeO₂<Ge-NCs> heterolayers used as a resist in laser lithography. We, however, believe that further considerable progress can be achieved along this line of research. However, in this way, even at the technological level available to us it was possible to fabricate some simple optical devices. In particular, the evaporation process of GeO₂<Ge-NCs> heterolayers by fs laser beam focused to a 1-μm diameter spot was used to fabricate prototypes of diffraction gratings (see Fig. 25). GeO₂<Ge-NCs> heterolayers 25 – 300 nm thick were scanned with laser beam at a speed 100 μm/s and frequency 1 kHz in the air. For the chosen heterolayer thickness, laser pulse energy was matched to slightly exceed the film evaporation energy threshold. The results showed that, under such conditions, it was possible to prepare diffraction gratings with line density up to 1000 lines per 1 mm.

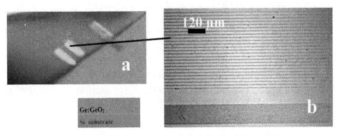

**Figure 25.** Diffraction of white light (**a**) on a set of lines in GeO₂<Ge-NCs> heterolayer on Si-substrate formed by a dry laser lithography (**b**) (optical microscopy photos).

As is seen, the main idea behind this method is that laser irradiation can be used to heat Ge-nanoparticles incorporated in a continuous medium. This idea differs little from the idea of using such Ge-nanoparticles in nanofoam formation from GeO₂<Ge-NCs> heterolayers. Therefore, we tried to employ Ge-nanoparticles in destructing malignant tumors. First, it is necessary to become able to prepare germanium particle based colloid solutions. It may be not so difficult a problem. A GeO₂ matrix with embedded Ge-nanoclusters is readily soluble in water or aqueous solutions of various active organic or inorganic chemicals. It is easy to find among the various chemicals those inert to germanium. Using such solutions, we can transfer Ge-nanoclusters from GeO₂<Ge-NCs> heterolayers into a suitable colloidal solution. This is one possible way towards solving the problem. Another strategy here is based on the ability of germanium to relatively slowly dissolve not only in liquid aggressive media, both acid and alkaline, but also in water. The solving proceeds in two stages: (*i*) oxidation of Ge-nanoparticle surface to GeO₂; (*ii*) dissolution of the oxide in a solvent. In some time, the colloidal germanium particles introduced in human organism will become completely dissolved to be removed from it in a natural way.

Another possible application of the results described above may be the use of Ge-NCs in medicine to suppress malevolent tumour formation. A method of tissue destruction in malignant tumours is widely known; it employs injection of colloidal solutions containing Ag-NCs with average diameter ~ 5-7 nm into tumor (Tyurnina et al., 2011). Ag-NCs spread over the tumour body and penetrate some part of its cells. Then the tumour is to be exposed to a laser beam which burns out the soft tumour tissues to a big depth due to heating of Ag-NCs.

## 5. Conclusions

The main aim of this paper was to describe the technological potential of pulsed laser treatments in modification of properties of film heterosystems involving germanium oxide layers with embedded Ge-nanoparticles. The goal of the study was the development of a general concept of physical-chemical mechanisms underlying such processes, i.e. suggestion 1 of a physical picture, or materials-science concept with analysis of reasons and outcomes, rather than the search for particular processes ensuring required modifications together with identification of appropriate technological conditions.

The general scientific aspect of the research involved two tasks. One of the tasks concerned fundamental problems in studying effects and processes connected with the input of large amounts of energy into small volumes of a solid, i.e. with the behavior of substance absorbing ultra-short laser pulses and subsequent relaxation of excitations in the solid body and surrounding medium. The second task was investigation of changes of structura chemical, physical and electrical properties of thin films under such treatments. New was the study of structural changes under pulsed laser treatments in multi-layer insulating coatings containing the films with a high concentration of QDs in the form of semiconductor nanoparticles. More specifically, we examined only one kind of such films, formed by germanium oxide layers (GeO and $GeO_2$) with embedded Ge-QDs.

A characteristic feature of both kinds of the effects due to PLA on studied multilayer coatings was that ultra-short laser pulses locally absorbed in small volumes of some cubic microns led to relaxation of this energy in volumes many times bigger than the sizes of the absorption zone while the duration of the dissipation process of absorbed energy lasted for a time that exceeded the laser pulse duration by many orders.

If it were not so, then we could not detect in the irradiated films indications of such slow and very large-scaled processes as films densification (by some ten percents), diffusion (up to 30% of the film material mass at distances of some hundred nm), rearrangement of the atomic network of the material from one kind into another, and ablation. In other words, all these modification processes of the material structure show that part of the material absorbed laser pulse energy from subsystems with superfast processes (subsystems of electromagnetic radiation, electron and phonon subsystems) spread into the subsystem of lattice processes. The latter subsystem has longest excitation relaxation times during chemical and structural rearrangement processes. Those long relaxation times are due to the fact that these rearrangements are collective, involving many atoms. Within such ideas, it can be speculated that, increasing the laser power, we will enter the conditions under which part of the energy absorbed by the solid will become sufficient for excitation of shock waves in the sample. These waves will begin to mechanically destroy the sample into fragments scattering around. This effect is called ablation.

An analysis of experimental data showed that, under PLA, layered films capable of good absorption of laser radiation play a big role in laser pulse energy accumulation. In the present case, those were GeO(solid) layers, GeO$_2$<Ge-NCs> heterolayers, and also the silicon substrate. All of the layers conveyed the absorbed energy either in heat or in structural chemical modification processes. The latter is clearly seen in Fig. 17 e and f, where the ablation process in the treated materials begins as the formation of ripples well before the thick SiO$_2$ layers covering the materials are exploded. In cases where thick GeO$_2$<Ge-NCs> heterolayers were absent from the multilayer coatings or when the silicon substrate absorbing radiation was replaced with transparent glass, increased threshold energies for all the processes activated in the multilayer system under PLA were observed.

From these standpoints, one can suggest an explanation to the opposite results obtained for two kinds of fs and ns PLA of GeO$_2$<Ge-NCs> heterostructures when, in one case, we had the formation process of super-loose nanofoam-type structure and, in the other case, a layer of a material having a very high density. In fs laser treated GeO$_2$<Ge-NCs> heterolayers, the laser pulses, 30 fs long, follow at 1-ms intervals, with the predominant part of their energy being absorbed by Ge-NCs. Suppose that this energy has enough time to completely dissipate for less than 1 msec. Then, strictly speaking, an act of energy absorption by this Ge-nanoparticle is no different from a similar act for the next pulse. This means that, in each of the pulses, identical portions of nanocluster-absorbed energy will produce identical amounts of gaseous monoxide at the boundary of the Ge-nanocluster with the surrounding GeO$_2$ glassy matrix. But to return the system to its strictly initial state till the coming of the next pulse, gas GeO formed around the Ge-nanoparticle should again transforms into initial solid components of Ge and glassy GeO$_2$. Otherwise, the absorption act of the second laser pulse by Ge-nanoparticle will not be strictly identical to the first one. It means that the whole nanofoam formation process will be confined to the process with a periodical alternation of two phases – one during which minimal setting nanofoam quantities form for less than 1 ms and the phase during which they disappear so quickly.

A different result was obtained in the experiment: nanofoam formation turned a stable and irreversible process. Logically, henceforth it follows that, in a series of fs laser pulses, each is absorbed by the GeO$_2$<Ge-NCs> heterolayer in the conditions different from the absorption of the previous ones. So, the effect of their impact should be considered not as a set of discreet and disconnected flashes, but as some continuous process proceeding during the duration of a whole pulse series. In fact, it shows that the complete time of energy relaxation by GeO$_2$<Ge-NCs> heterolayer absorbed from a separate pulse is longer than the intervals between them. Therefore, accumulation of some energy part of each pulse proceeds in the heterolayer. This accumulation integrates the whole series of discreet flashes in a continuous action that consists of three stages – the initial stage, the stage of stationary mode and the final stage that proceeds upon the completion of radiation. During PLA of GeO$_2$<Ge-NCs> heterolayer with ns laser, the time spans between laser pulses are so big that one cannot say about the thing that their effect is connected with part of their energy accumulation by GeO$_2$<Ge-NCs> heterolayer from pulse to pulse. Besides, GeO$_2$ matrix considerably absorbs ns laser pulse energy in the

$GeO_2$<Ge-NCs> heterolayer. Accordingly, the part of the pulse energy absorbed by Ge-nanoparticles considerably decreases and, along with it, their heating rate also decreases. If the speed of Ge-NCs heating in the heterolayer is lower than that of $GeO_2$ matrix, then the matrix will be cool down close to the border between these two heterolayer components, as it spends part of its heat for heating colder Ge-NCs. In the rest part of the glassy $GeO_2$ matrix, with the growing temperature, the decay processes of atomic net remaining parts of metastable GeO(solid) and its shrinkage by lowering atomic $GeO_2$ net defectiveness proceed. The latter of these processes is usually accompanied by a viscosity increase in glasses and that is to cause threshold energy growth of a beginning of foam formation in the $GeO_2$<Ge-NCs> heterolayer. Thus, the totality of all the described factors impairs, in this case, $GeO_2$<Ge-NCs> heterolayers transformation into a nanofoam-like matter under their ns laser pulse radiation under the conditions used in our experiments.

At the end of this research work, we emphasise one more time that it is difficult so far to find out of a multitude of film coatings and materials this or that way involved in microelectronics such their kinds that would characterised by so high capability to properties and structure modification and absolutely all the properties as germanium oxide layers. Although, as we believe, silicon oxide-based films are the most chemically close to them. These materials also have the capability to different modification forms and radical changes of many physicochemical properties. Particularly, in our viewpoint, $SiO_2$ layers transformation into nanofoam-like material may be one of the most interesting out of modifications similar to those of germanium oxides. It is not excluded that it will be realised by the way analogous to the used one for nanofoam formation from glassy $GeO_2$.

The authors hope that the possibilities demonstrated in our paper for germanium oxide layers modification assisted with laser treatments will attract attention of material scientists in the field of film coatings used in nano- and optoelectronics and also of researchers engaged in applied trends of nanotechnology.

## Author details

Evgenii Gorokhov, Kseniya Astankova, Alexander Komonov and Arseniy Kuznetsov
*Institute of Semiconductor Physics of SB RAS, Russia*
*Laser Zentrum Hannover, Germany*

## Acknowledgement

This investigation was supported by the Russian Fundation for Basic Research (projects Nos. 07-08-00438 and 10-07-00537). E.B. Gorokhov is grateful to Universite' de Nancy (France) and Laser Zentrum Hannover (Germany) for a visit grant. Also, the authors thank their colleagues for assistance in work: Dr. D.V. Marin, Dr. T.A. Gavrilova, Dr. V.A. Volodin, M. Slabuka, Prof. Boris Chichkov from Laser Zentrum Hannover (Germany), Prof. M. Vergnat from Institute Jean Lamaur and l'Universite' de Lorraine for their help in studying GeO(solid) films and Prof. A.V. Latyshev for financial support.

# 6. References

Ahmanov, S.A.; Yemel'yanov, V.I.; Koroteev, N.I. Seminogov, V.N. (1985). The Impact of High-Power Laser Irradiation on the Surface of Semiconductors and Metals: Nonlinear Optical Effects and Nonlinear Optical Diagnostics. *Advances in Physical Sciences*, Vol. 147, No. 4, pp. 675-745

Appen, A.A. (1974). Chemistry of Glass. 2nd Edition. *Publ.: Khimiya, Leningrad (In Russian)*.

Ardyanian, M.; Rinnert, H.; Devaux, X. & Vergnat, M. (2006). Structure and Photoluminescence Properties of Evaporated GeOₓ Thin Films. *Appl. Phys. Lett.*, Vol.89, pp. 011902-1-3

Ashitkov, S.I.; Ovchinnikov, A.V. & Agranat, M.B. (2004). Recombination of an Electron-hole Plasma in Silicon Under the Action of Femtosecond Laser Pulses. *JETP Letters*, Vol.79, No 11, pp. 529-531

Bok, J. (1981). Effect of electron-hole pairs on the melting of silicon. *Phys. Lett.*, Vol. 84 A., No. 8., pp. 448-450

Chong, T.C.; Hong, M.H. & Shi, L.P. (2010) Laser Precision Engineering: From Microfabrication to Nanoprocessing. *Laser&Photon. Rev.*, Vol.4, No 1, pp. 123-143

Dvurechenskiy, A.V.; Kachurin, G.A.; Nidaev, N.V.; Smirnov, L.S. (1982). Pulsed Annealing of Semiconductor Materials. *Publ.: Nauka, Novosibirsk (In Russian)*.

Ekimov, A.I. & Onushchenko, A.A. (1981). Quantum-Size Effect in Three-dimensional Microscopic Semiconductor Crystals. *JETP Lett.*, Vol.34, No.6, pp. 345-348

Filipovich, V.N. (1978). The Theory of Self-Diffusion of Oxygen in Glassy SiO₂, GeO₂. *Glass Physics and Chemistry*. Vol.4, No.1, pp. 22-30

Gallas, B.; Kao, C.-C.; Fisson, S.; Vuye, G.; Rivory, J.; Bernard, Y. & Belouet, C. (2002). Laser Annealing of SiOₓ Thin Films. *Appl. Surf. Science,* Vol. 185, pp. 317–320

Gorokhov, E.B.; Kosulina, I.G.; Pokrovskaya, S.V. & Neizvestny, I.G. (1987). Mechanical and Electrical Properties of the Double-layer Film System GeO₂-Si₃N₄ on Ge. *Phys. Stat. Sol. (a)*, Vol.101, pp. 451-462

Gorokhov, E.B.; Noskov, A.G.; Sokolova, G.A.; Stenin, S.I.; Trukhanov, E.M (1982). Mechanical Stability of Pyrolytic Silicon Dioxide Films. *Physics, Chemistry and Mechanics of Surfaces (In Russian – Surface)*, No.2, pp.25-33

Gorokhov, E.B.; Prinz, V.Y.; Noskov, A.G. & Gavrilova, T.A. (1998). A Novel Nanolithographic Concept Using Crack-assisted Patterning and Self Alignment Technology. *J. Electrochem. Soc.*, Vol.145, No.6, pp. 2120-2131

Gorokhov, E.B. (2005). *Evaporation and Crystallisation Processes in Germanium Oxide Films on Germanium*. PhD thesis, SB RAS, Novosibirsk, Russia

Gorokhov, E.B.; Volodin, V.A.; Kuznetsov A.I.; Chichkov B.N.; Astankova, K.N. & Azarov, I.A. (2011). Laser Treatment of the Heterolayers "GeO₂:Ge-QDs". *Proc. of SPIE*, Vol.7994, pp. 79940W-10

Hrubesh, L.W. & Poco, J.F. (1995). Thin Aerogel Films for Optical, Thermal, Acoustic and Electronic Applications. *Journal of Non-Crystalline Solids*, Vol. 188, pp. 46-53

Jambois, O.; Rinnert, H.; Devaux, X. & Vergnat M. (2006). Influence of the Annealing Treatments on the Luminescence Properties of SiO/SiO$_2$ Multilayers. *J. Appl. Phys.*, Vol.100, pp. 123504-6

Jolly, M. & Latimer, W.M. (1952). The Equilibrium Ge(s) + GeO$_2$(s) = 2GeO(gas). The Heat of Formation of Germanic Oxide. *J. Amer. Chem. Soc.*, Vol.74, No20, pp. 5757-5758

Juodkazis, S.; Mizeikis, V. & Misawa, H. (2009). Three-dimensional Microfabrication of Materials by Femtosecond Lasers for Photonics Applications. *J. Appl. Phys.*, Vol. 106. pp. 051101-1-14 Kachurin, G.A.; Pridachin, N.B.; Smirnov, L.S. (1975). Annealing of Radiation Defects by Pulsed Laser Irradiation. *Semiconductors*, Vol. 9, No. 7, pp. 1428-1429

Kamata, Y. (2008). High-k/Ge MOSFETs for Future Nanoelectronics. *Materials Today*, Vol.11, No.1-2, pp. 30-38, ISSN 1369-7021

Knoss, R.W. (2008). *Quantum Dots: Research, Technology and Applications*, Nova Science Publishers Inc., New York

Korchagina, T. T.; Gutakovsky, A. K.; Fedina, L. I.; Neklyudova, M. A. & Volodin, V. A. (2012). Crystallisation of Amorphous Si Nanoclusters in SiO$_x$ Films Using Femtosecond Laser Pulse Annealings. *Journal of Nanoscience and Nanotechnology* (in print)

Korte, F.; Nolte, S.; Chichkov, B.N.; Bauer, T.; Kamlage, G.; Wagner, T.; Fallnich, C. & Welling H. (1999). Far-field and near-field Material Processing with Femtosecond Laser Pulses. *Appl. Phys. A.*, Vol. 69, pp. S7-S11

Marin, D.V.; Gorokhov, E.B.; Borisov, A.G. & Volodin, V.A. (2009). Ellipsometry of GeO$_2$ Films with Ge Nanoclusters: Influence of the Quantum-size Effect on Refractive Index. *Optics and Spectroscopy*, Vol.106, No3, pp. 436-440, ISSN 0030-400X

Marin, D.V.; Volodin, V.A.; Gorokhov, E.B.; Shcheglov, D.V.; Latyshev, A.V.; Vergnat, M.; Koch, J. & Chichkov, B.N. (2010). Modification of Germanium Nanoclusters in GeOx Films During Isochronous Furnace and Pulse Laser Annealing. *Technical Physics Letters*, Vol.36, No5, pp. 439-442

Martynenko, A.P.; Krikorov, V.S.; Strizhkov, B.V. & Marin, K.G. (1973). Physicochemical Properties of Silicon and Germanium Monoxide. *Inorganic materials*, Vol.9, No.9, pp. 1394-1399

Martsinovsky, G.A.; Shandybina, G.D.; Dement'eva, Yu.S.; Dyukin, R.V.; Zabotnov, S.V.; Golovan', L.A. & Kashkarov, P.K. (2009). Generation of Surface Electromagnetic Waves in Semiconductors Under the Action of Femtosecond Laser Pulses. *Semiconductors*, Vol.43, No10, pp. 1298-1304

Molinari, M.; Rinnert, H. & Vergnat, M. (2003). Visible Photoluminescence in Amorphous SiOx Thin Films Prepared by Silicon Evaporation Under a Molecular Oxygen Atmosphere. *Appl. Phys. Lett.*, Vol.82, No.22, pp. 3877-3879

Mueller, R.L. (1960). Chemical Characteristics of Polymer Glass-Forming Substances and the Nature of Glass Formation. Book: Proceedings of the Third All-Union Conference on the Glassy State. *Publ.: The USSR Academy of Sciences, Moscow and Leningrad.* Pp.61-71

Mueller, R.L. (1955). A Valence Theory of Viscosity and Fluidity for High-Melting Glass-Forming Materials in the Critical Temperature Range. *Journal of Applied Chemistry (Russian Journal of Applied Chemistry).* Vol.28, No.10, pp. 1077-1087

Nelin, G. & Nilsson, G. (1972). Phonon Density of States in Germanium at 80 K Measured by Neutron Spectrometry. *Phys. Rev. B*, Vol.5, pp. 3151-3160

Nemilov, S.V. (1978). The Nature of the Viscous Flow of Glasses with Frozen Structure and Some of the Consequences of Valence-Configuration Theory of Viscous Flow. *Glass Physics and Chemistry*. Vol.4, No.6, pp. 662-674

Nemilov, S.V. (1978). The Valence-Configuration Theory of Viscous Flow of Supercooled Glass-Forming Liquids and its Experimental Validation. *Glass Physics and Chemistry*. Vol.4, No.2, pp. 129-148

Ogden, J.S. & Ricks, M.J. (1970). Matrix Isolation Studies of Group IV Oxides. II. Infrared Spectra and Structures of GeO, Ge$_2$O$_2$, Ge$_3$O$_3$, and Ge$_4$O$_4$. *J. Chem. Phys.*, Vol.52, No1, pp. 352-357

Pliskin, W.A.; Lehman, H.S. (1965). Structural Evaluation of Silicon Oxide Films. *J. Electrochem. Soc.*, Vol.122, pp.1013-1019

Rinnert, H.; Vergnat, M.; Burneau, A. (2001). Evidence of Light-Emitting Amorphous Silicon Clusters Confined in a Silicon Oxide Matrix. *J. Appl. Phys.* Vol. 89, No. 1, pp. 237-243

Rochet, F.; Dufour, G.; Roulet, H.; Pelloie, B.; Perrier, J.; Fogarassy, E.; Slaoui, A. & Froment M. (1988). Modification of SiO Through Room-temperature Plasma Treatments, Rapid Thermal Annealing, and Laser Irradiation in a Nonoxidizing Atmosphere. *Phys. Rev. B.*, Vol.37, No 11, pp. 6468-6477

Salihoglu, O.; Kürüm, U.; Yaglioglu, H.G., Elmali, A.; Aydinli, A. (2011). Femtosecond laser crystallization of amorphous Ge. *J. Appl. Rhys.* Vol.109, pp. 123108-1

Sameshima, T. & Usui, S. (1991). Pulsed Laser-induced Amorphization of Silicon Films. *J. Appl. Phys.*, Vol. 70. No. 3, pp. 1281-1289

Sanditov, D.S. (1976). On the Mechanism of Viscous Flow of Glasses. *Glass Physics and Chemistry*. Vol.2, No.6, pp. 515-518

Sheglov, D.V.; Gorokhov, E.B.; Volodin, V.A.; Astankova, K.N. & Latyshev, A.V. (2008). A Novel Tip-induced Local Electrical Decomposition Method for Thin GeO Films Nanostructuring. *Nanotechnology*, Vol.19, pp. 245302-1-4

Takeoka, S.; Fujii, M.; Hayashi, S. & Yamamoto, K. (1998). Size-dependent Near-infrared Photoluminescence From Ge Nanocrystals Embedded in SiO$_2$ Matrices. *Phys. Rev. B.*, Vol.58, pp. 7921-7925

Tananaev, I. V.; Shpirt, M. Ya. (1967). The Chemistry of Germanium. Khimiya, Moscow (in Russian).

Tyurnina, A.E.; Shur, V.Ya.; Kuznetsov, D.K.; Mingaliev, E.A. & Kozin, R.V. (2011). Synthesis of Silver Nanoparticles by Laser Ablation in Liquid, *Proceedings of 19th International Symposium "Nanostructures: Physics and Technology"*, pp. 111-112, ISBN 978-5-93634-042-0, Ekaterinburg, Russia, June 20-25, 2011

Volodin, V.A.; Efremov, M.D; Gritsenko, V.A. & Kochubei S. A. (1998). Raman Study of Silicon Nanocrystals Formed in SiN$_x$ Films by Excimer Laser or Thermal Annealing. *J. Appl. Phys. Lett.*, Vol.73, pp. 1212–1214

Volodin, V.A.; Gorokhov, E.B.; Marin, D.V.; Cherkov, A.G.; Gutakovskii, A.K. & Efremov, M.D. (2005). Ge Nanoclusters in GeO$_2$: Synthesis and Optical Properties. *Solid State Phenomena*, Vol.108-109, pp. 83-90

Volodin, V.A.; Korchagina, T.T.; Kamaev, G.N.; Antonenko, A.Kh.; Koch, J.; Chichkov, B.N.
    (2010). Femtosecond and Nanosecond Laser Assistant Formation of Si Nanoclusters in
    Silicon-Rich Nitride Films. *Proceedings of International Conference "Micro- and
    Nanoelectronics"*, pp. 75210X1-(X8), SPIE, Vol. 7521, Zvenigorod, Moscow region, Russia,
    October 5-9, 2009.
Zabotnov, S.V.; Golovan', L.A.; Ostapenko, I.A.; Ryabchikov, Yu.V.; Chervyakov, A.V.;
    Timoshenko, V.Yu.;   Kashkarov, P.K. &   Yakovlev, V.V. (2006). Femtosecond
    Nanostructuring of Silicon Surfaces. *JETP Letters*, Vol. 83, No. 2, pp. 69-71
Zakis, Yu.R. (1981). Applicability of Ideas about *Quasi-Particles and Defects* to *Glass*. *Glass
    Physics and Chemistry*. Vol.7, No.4, pp. 385-390

# Chapter 4

# Jet Engine Based Mobile Gas Dynamic $CO_2$ Laser for Water Surface Cleaning

V. V. Apollonov

Additional information is available at the end of the chapter

## 1. Introduction

What is the best to start our paper with? "Deep water Horizon" (BP's operation) case in the Mexican gulf is the best example of man made natural disasters. Today, the notion that "offshore drilling is safe" seems absurd. The Gulf spill harks back to drilling disasters from few decades past — including one off the coast of Santa Barbara, Calif. in 1969 that dumped three million gallons into coastal waters and led to the current moratorium. The "Deep water Horizon" disaster is a classic "low probability, high impact event" — the kind we've seen more than our share of recently, including space shuttle disasters, 9/11, Hurricane Katrina and earth quake in Japan. And if there's a single lesson from those disparate catastrophes, it's that pre-disaster assumptions tend to be dramatically off-base, and the worst-case scenarios downplayed or ignored. The Gulf spill is no exception. Fire boats battle the fire on the oil rig "Deep water Horizon" after the April 21 terrible explosion.

The post-mortem is only the beginning, so the precise causes of the initial explosion on the drilling platform and the failure of a "blowout preventer" to deploy on the sea floor probably won't be established for weeks or months. But the outlines of serious systemic problems have already emerged, indicating just how illusory the notion of risk-free drilling really was, while pointing to some possible areas for reform. A "blowout" on an oil rig occurs when some combination of pressurized natural gas, oil, mud, and water escapes from a well, shoots up the drill pipe to the surface, expands and ignites. Wells are equipped with structures called blowout preventers that sit on the wellhead and are supposed to shut off that flow and tamp the well. "Deep water Horizon"'s blowout preventer failed. Two switches — one manual and an automatic backup — failed to start it. When such catastrophic mechanical failures happen, they're almost always traced to flaws in the broader system: the workers on the platform, the corporate hierarchies they work for, and the government bureaucracies that oversee what they do. According to the study 600 major equipment failures in offshore drilling structures 80 percent were due to "human and organizational factors", and 50 percent of those due to flaws in the engineering design of equipment or processes. With near-shore and shallow reserves of fossil fuels largely depleted, drilling has moved farther off shore, into deeper waters and deeper underground. The technology for locating oil and gas reserves and for drilling has improved, but the conditions are extreme and the challenges more formidable. This is a pretty frigging complex system. You've got equipment and steel strung out over a long piece of geography starting at surface and terminating at 18,000 feet below the sea floor. So it has many potential weak points. Just as Katrina's storm surge found weaknesses in those piles of dirt — the levees — gas likes to find weakness in anything we connect to that source. It must be questioned, whether energy companies and government agencies have fully adapted to the new realities. The danger has escalated exponentially. We've pushed it to the bloody edge in this very, very unforgiving environment, and we don't have a lot of experience. Disaster has several possible insights for the oil spill: one was that BP and other corporations sometimes marginalize their health, safety, and environmental departments. BP and other companies tend to measure safety and environmental compliance on a day-to-day, checklist basis, to the point of basing executive bonuses on those metrics. But even if worker accident rates fall to zero, that may reveal nothing about the risk of a major disaster. These things we are talking about are risks that won't show up this year, next year — it may be 10 years down the road before you see one of these big blowouts or refinery accidents.

That assumption — that catastrophic risks were so unlikely they were unworthy of serious attention — appears to have driven a lot of the government decision-making on drilling as well. One, published in 2007, estimated the "most likely size" of an offshore spill at 4,6 K barrels. Current, conservative estimate of the Gulf spill put its total at more than 80 K barrels, increasing at a rate of 5 K per day. Why we have paid in our paper such a detailed attention to the policy of BP`s "Deep water Horizon"? The contamination of large water areas (oceans, seas, lakes, and rivers) with petroleum products as a result of accidents and not so smart industrial activity of similar companies and officials around them is one of major problems of protection of environment. Any other contaminant can not be compared with petroleum on a basis of universal utilization, number of contamination sources and degree of effect on all components of environment. Through penetration of petroleum products into water, there are deep, frequently irreversible changes of its chemical, physical and microbiological and even global (redirection of the ocean streams) properties.

The following methods of disposal of petroleum contamination of water: natural, mechanical, physical-chemical, chemical, biochemicals are widely used nowadays.

Natural method demands to leave the oil alone so that it breaks down by natural means. If there is no possibility of the oil polluting coastal regions or marine industries, the best method is to leave it to disperse by natural means. A combination of wind, sun, current, and wave action will rapidly disperse and evaporate most oils. Light oils will disperse more quickly than heavy oils.

The mechanical methods are such: collecting of petroleum from a surface manually or with different installations. All these methods are effective during limited time (from several hours to several days) only – time period during which the thickness of a petroleum film is great enough. Contain the spill with booms and collect it from the water surface using skimmer equipment. Spilt oil floats on water and initially forms a slick that is a few millimeters thick. There are various types of booms that can be used either to surround and isolate a slick, or to block the passage of a slick to vulnerable areas such as the intake of a desalination plant or fish-farm pens or other sensitive locations. Boom types vary from inflatable neoprene tubes to solid, but buoyant material. Most rise up about a meter above the water line. Some are designed to sit flush on tidal flats while others are applicable to deeper water and have skirts which hang down about a meter below the waterline. Skimmers float across the top of the slick contained within the boom and suck or scoop the oil into storage tanks on nearby vessels or on the shore. However, booms and skimmers are less effective when deployed in high winds and high seas.

Use dispersants to break up the oil and speed its natural biodegradation. Dispersants act by reducing the surface tension that stops oil and water from mixing. Small droplets of oil are then formed, which helps promote rapid dilution of the oil by water movements. The formation of droplets also increases the oil surface area, thus increasing the exposure to natural evaporation and bacterial action. Dispersants are most effective when used within an hour or two of the initial spill. However, they are not appropriate for all oils and all locations. Successful dispersion of oil through the water column can affect marine organisms like deep-water corals and sea grass. It can also cause oil to be temporarily accumulated by sub-tidal seafood. Decisions on whether or not to use dispersants to combat an oil spill must be made in each individual case. The decision will take into account the time since the spill, the weather conditions, the particular environment involved, and the type of oil that has been spilt and many other parameters involved into that hard consideration.

Introduce biological agents to the spill to hasten biodegradation. Most of the components of oil washed up along a shoreline can be broken down by bacteria and other microorganisms into harmless substances such as fatty acids and carbon dioxide. This action is called biodegradation. The natural process can be speeded up by the addition of fertilizing nutrients like nitrogen and phosphorous, which stimulate growth of the microorganisms concerned. However the effectiveness of this technique depends on factors such as whether the ground treated has sand or pebbles and whether the fertilizer is water soluble or applied in pellet or liquid form.

Besides of that the biological activity of microorganisms strongly depends on the temperature of water. As for physical-chemical methods it is necessary to mention, first of all, application of various adsorbing materials (polyurethane foam, coal dust, sawdust etc.), however all these methods are labor-consuming and low efficient. Besides they require secondary reprocessing of adsorbents.

Chemical method is the method of petroleum removal with the help of chemical substances. The basic disadvantages of this method are high price and fact, that detergents frequently are more toxic for water microorganisms, than petroleum.

The laser method of cleaning of water surface from a thin petroleum film is one of physical-chemical methods. Some time later after penetration into water, the petroleum spreads on a surface of a water and forms very thin film (thickness several microns). This film cannot be forced to burn, since because of a good thermal contact to a surface of water the film cannot be heated to temperature ensuring steady combustion.

The principle idea of a laser method consists in following. The laser beam passes through a petroleum film and then is absorbed in thin layer of water. The water heats very fast up to the boiling temperature, and the forming vapor destroys a film, bursting it in small-sized fragments, which, mixing up with the hot vapor, are decomposed quickly with formation of simple un-toxic substances.

The main advantages of a laser method consist of the following.

a.   This method is "fast-response" one, since does not require any special preparation; in emergency situations the time from the moment of obtaining of the alarm signal to the beginning the laser installation operation is determined only by time necessary for arrival of the ship or the helicopter with the installation on board in given area.

b.   The method is contactless, i.e. does not require realization of preparatory or other activities in the oil spillage.

For realization of the proposed technology, it is possible to utilize different types of lasers, as continuous (power up to 250 kW), and high repetition rate pulse-periodical (average power of the same level with duration of pulses 100 ns and repetition rate up to 100kHz).

Theoretical estimations and the experiments have shown, that as a result of a laser beam action on a surface of a water and land, covered with film of hydrocarbon contamination, following effects can be observed:

Evaporation and burning, and in a continuous mode the consumption of energy for 1 gram of vaporized liquid on the order of value surpasses energy necessary for heating up to boiling temperature and evaporation of a film, that is explained by a heat consumption for heating of water. In a pulse mode, the consumption of energy it is per unit of vaporized mass of a liquid (film) approximately in 5 times less, thus the process of film ignition start more easily.

Knocking out the particles of polluting substance above a surface of water under action of pulse or powerful scanning laser radiation should be considered [1]. The physical mechanism of this effect is explained by sharp evaporation (boiling-up) of a thin layer of water under a layer of polluting substance. This process takes place where as hydrocarbon polluting thin film has an absorption coefficient less, than water. Knocked out particles of petroleum at a height up to 50 cm is possible. For this case, the energy consumption per unit mass of a raised liquid is sufficiently less than energy necessary for evaporation and ignition (mode number one) or for sucking up and saving for future efficient usage (mode number two).

The character of a task to be solved superimposes certain conditions on operating characteristics of the laser device concerning both parameters of laser source, and concerning auxiliary systems.

Technological lasers of 10 - 15 kW power range, that widely are used in production, will have the output completely insufficient for liquidation of large scale contamination presenting the greatest danger according to the above-stated estimations. Such lasers can be utilized, at the best, for improvement of a process engineering of cleaning under modeling conditions.

Besides, the character of this task dictates impossibility or extreme undesirability of use of the stationary civil engineering service line (electric power network, water pipe, main gas line etc.), in view of the requirement of a high self-sufficiency and mobility. The power unit should not limit mobility and thus provide totally energy needed of all installation, the necessary reserve of expendables and fuel on board a complex. A capability of a fast redeployment from one type of a vehicle on another is also desirable.

In the present technical paper, the basic design concept of the mobile laser installation on basis of gas dynamic $CO_2$-laser is developed. The activities were executed in accordance with the working plan of OOO "Energomashtechnika"

In the first paragraph, the possibility of using of various types of lasers for solution of required tasks is considered and the selection of GDL is justified.

The second paragraph is devoted to selection and substantiation of basic performances of the laser installation and principal schematics of GDL.

In the third paragraph, the possible schemes of organization of air supply in the laser installation are considered. The selection of jet engines for high quality power unit of GDL installation has been provided. The necessary volume of the selected jet engine adaptation/modification experimental works is presented.

The fourth paragraph of the paper is devoted to the description of the design concept of GDL, and also laser installation as a whole. The pneumohydraulic schematics are described, the structure of the equipment, fuel tanks and operation control units are determined. The description of jet engine based CO2 GDL and detailed analysis of the laser installation components is also presented.

## 2. The substantiation of selection of laser type according to specific parameters, operation autonomy, and mobility

There are few the most effective and scalable modern high power continuous/pulse-periodical operated gas lasers should be considered for realization of clamed in the title of that paper tasks: electro discharge laser (EDL), gas dynamic laser (GDL), chemical laser

HF/DF (CL) and chemical oxygen - iodine laser (COIL).We are not going here to details of high power high repetition rate P-P laser systems operation.Main results of oil films elimination detailed consideration taken in the past can be summarized and reduced to the paper format and presented here as following:

## 2.1. Analysis of applicability of various types of lasers to the task

The laser with output power up to 250 kW of continuous operation (CW) or high repetition rate pulse-periodical operation (P-P) during few hours (minimum requirement) is required for realization of the mentioned above task. Besides of that, according to the physics of the process of destruction of a petroleum film, explained in Introduction, laser radiation should be weakly absorbed by petroleum film and should be absorbed effectively by water. In the table 1, the computational data for depth of radiation penetration for 4 types of mentioned above lasers in petroleum and in water are presented.

| The laser type | Depth of penetration, micron | |
|---|---|---|
| | Petroleum | Water |
| GDL, EDL | 100 - 260 | 10 |
| CL | 26 - 200 | 0,8 - 40 |
| COIL | 50 | 2000 |

**Table 1.** Laser radiation penetration depth for petroleum and water.

The range of values of depth of penetration for petroleum is the sequence of the fact, that absorption coefficient of various grades of petroleum and water (for the CL case) in relation to wavelength in CL radiation spectral band is considered.

From the Table 1, it is obvious, that the COIL radiation is rather weakly absorbed by water and is rather strongly absorbed by petroleum. However to heat a petroleum film by thickness ~100 microns through the absorption mechanism of COIL radiation practically is impossible because of strong heat transfer from a film into water through the heat conductivity. The conclusion is: the use of COIL to solve this task is impossible.

As the lengths of waves of radiation GDL and EDL are identical, the comparison of these two types of lasers is carried out on the basis of mass - dimension characteristics and other parameters that are listed in Table 2. The comparison is carried out for two real installations, which were developed and tested in the past to sufficient degree, precisely - GDL of rated power up to 250 kW, and EDL with 20kW output power.

The analysis of the data, presented in the Table 2 shows, that GDL has decisive advantages in comparison with EDL. It is necessary to be mentioned here, that EDL specific fuel consumption is 2 times less, than that of GDL. However, influence of this factor on the total complex weight will demonstrate a negative influence only for large operation time (5 hours or more). At the arrangement of a laser complex on the board of helicopter, the operating time will not exceed 1 hour, therefore in this case smaller EDL specific fuel consumption will not take decisive value.

| #[1] | The name | GDL | EDL |
|------|----------|-----|-----|
| 1. | Power laser (max), kW | Up to 250 | 20 |
| 2. | Overall dimensions of the laser installation, m | 4x2.4x2.4 | 2.5x3x3.5 + 2 x 0,8 x 0,8 |
| 3. | "Dry weight " (free of fuel), metric ton | 5 | 18.6 |
| 4. | Specific, 1 kW of laser power dry weight, t/kW | Up to 0,02 | 0,93 |
| 5. | Specific, 1 kW of laser power, dry volume, cubic m/kW | 0,1 | 1,5 |
| 6 | Specific volume fuel consumption for 1 kW of laser power, cubical meter/ hour / kW | 0,02 | 0,01 |

**Table 2.** The comparison of the parameters for GDL (250 kW) and EDL (20 kW).

Thus, GDL and CL are remaining the only competing systems. The comparative analysis of these two types of lasers with reference to the considered task is stated below.

### 2.1.1. Specific power of laser generation

GDL specific power, Wsp is about 20-35 J/g (C2H4+toluene); this value for CL is much higher and reaches 150 J/g. From this point of view, CL has big advantage in comparison with GDL. But taking into account few other parameters one can say it is not the final conclusion for this particular story. Very important parameters for our task to be solved should be considered as well: wavelength, scalability of the system, technical maturity of the technology, safety, life time of hardware and so on. Up to now GDL looks like the best system for the task under consideration - cleaning of water surface from petroleum films.

### 2.1.2. Wavelength of radiation

GDL radiation wavelength, $\lambda = 10,6$ microns and CL wavelength, $\lambda = 2,7 - 4,5$ microns. From this point of view, CL has one more potential advantage over GDL, as the diffraction limited angle of divergence of CL beam is 3 - 4 times less in comparison with GDL (at identical aperture). However this problem requires the further presize consideration.

a. Radiation extraction from the resonator.

In high-power lasers, extraction of laser beam from the resonator, where the pressure is much lower than atmospheric pressure, in external space through a rigid window is practically impossible for the reason of thermal destruction of such a window. For a extraction of such a beam out of installation, the aerodynamic window (AW) usually can be utilized. The supersonic gas stream cross-sectional to the beam direction fulfills the task of transparent boundary. As there are areas of various density gas sheaths limited by curvilinear surfaces in such flow, the beam transmission through this gas inuniformity introduces the distortions of a wave front resulting in increase of an angular divergence of a beam. These distortions do not depend on a wavelength, as the dispersion of an index of refraction of gas is extremely small and therefore these distortions have the same absolute value for GDL and CL. However relative values of these distortions of a wave front set (i.e. reduced to wavelength)   are for CL 3-4 times higher, than for GDL, therefore from the point of view of an angular divergence of radiation the advantage CL in comparison with GDL can be not so essential.

b.    External optical system.

The optical systems GDL and CL for delivery beam energy to remote objects include necessarily external (in relation to the resonator) optical elements and systems, intended for expansion of a beam with the purpose of reduction of its divergence. A principal component of a telescope is the main mirror, which diameter can reach the value up to several meters (depending on required power density on the target). The one piece glass or metal mirror will be too heavy. For this reason the main mirror is usually considered as consisting from large number of small mirrors (facets), assembled in one unit. Every facet has the own adjusting device. The accuracy of relative positioning these facets should be not worse $\lambda/4$, otherwise distortions of a wave front after reflection of a laser beam from this mirror will cause the essential increase of an angular divergence in comparison with it diffraction limit. This is $\lambda/4 \cong 2,5$ microns for GDL , $\lambda/4 \cong 0,75$ microns for CL. Therefore in practice required accuracy is much more difficult to ensure for CL.

This brief discussion shows, that the efficiency of CL external optical system can be much below, than that of GDL and, hence, reduces advantage of shorter wavelength of CL radiation.

Besides of that, for small delivery length of the beam energy down to the water surface, the angular divergence has no great importance, since density of power on a water surface appears sufficiently high without application of an additional external optical system.

### 2.1.3. Length of radiation generation zone

In GDL working section (the resonator region) the transfer of oscillation energy from $N_2$ to $CO_2$ has taken place. The characteristic time of this process is relatively large. Besides the relaxation time of oscillatory - exited molecules $CO_2$ at their collisions with other particles is also large. Therefore for full extraction of $N_2$ accumulated oscillatory energy, and transformation it in laser radiation, length of a working zone in flow down direction should be 10 - 15 cm. Such large length provides certain freedom of selection of the resonator scheme (one pass or multipass, symmetrical or asymmetrical, etc.). Thus the mirror operating surfaces are large enough, that results in a reduction of its thermal loading and distortion of a surface due to inhomogeneous heating.

In CL, the length of beam generation section is small, less then 2 cm. This circumstance causes difficulties in optimization of the laser resonator and leads to increase of thermal loading of mirrors, which results to significant thermal distortion of mirrors and distortion of a wave front, i.e. to significant increase of an angular divergence of laser radiation.

### 2.1.4. Direct exhaust of used gas to atmosphere

If GDL or CL are placed on ground or on the board of flying vehicle (plane or helicopter), there is a problem of exhaust of spent gas to atmosphere. In the case of GDL this problem can be solved rather simply by use of the diffuser, as the operating pressure in a zone of the resonator is rather high. For reaching a maximum degree of pressure recovery in the diffuser it should have a very special geometry.

In the CL case the operating pressure in the resonator is much lower in comparison with GDL, therefore the direct exhaust to atmosphere with the help of the diffuser is impossible, it is necessary to utilize additionally gas or water vapor ejector to ensure necessary pressure in output section of the diffuser. Such ejector requires large additional gas flow rate, that results in fundamental complexity of the laser installation design and significantly decrease specific power of generation, i.e. in this case one of main CL advantages is lost in comparison with GDL.

### 2.1.5. Toxic characteristics

The degree of danger CL from the point of view of toxic characteristics of working components and exhaust gases is much higher in comparison with GDL. In CL, for creation of a working mixture the extremely toxic substances containing fluorine should be utilized. Besides the exhaust CL contains a significant quantity of fluorine hydride HF, which is supertoxicant. In this connection with use CL in a system of laser cleaning of water areas it will be necessary to ensure special safety measures, that will cause the significant complication of the installation design and rise of its price.

In the case of GDL exhaust gas is ecologically clean ($N_2 + CO_2 + H_2O$). However, CO, carbon monoxide which is toxic gas can be used as a fuel for GDL. Hence, system of storage and supply of CO should ensure absence of its leakage. On the other hand, CO is lighter-than-air, therefore it floats in atmosphere and is fast blown away by wind. If as fuel in GDL will be used liquid hydrocarbon fuel (benzole, toluene), the system of its storage and supply does not require special measures to secure the ecological safety.

### 2.1.6. Absorption of laser radiation by atmosphere

The wavelengths of GDL and CL radiation are within of windows of atmosphere transparency. These windows are usually determined for vertical transmission of beam through atmosphere, and for this case power dissipation on track is ~20 % for GDL and 25 % for CL. For vertical tracks of length 50 - 150 m, which are of interest with reference to a considered problem, the radiation power dissipation can be neglected.

## 2.2. Substantiation of selection gas dynamic $CO_2$-laser

Proceeding from the comparison of various type lasers, explained in the previous paragraphs, it is possible to make a conclusion, that the most suitable laser for development of the installation for disposal of petroleum films is GDL:

### 2.2.1. Simplicity of design

From design point of view GDL components are the most simple in comparison with the considered lasers. The combustion chamber operates on usual components, for example, kerosene + air. The design of such chambers is practically very much mature, there is a wide experience of their operation in various technical devices. The nozzle unit of GDL is made from heat resisting steel and can work long time without forced cooling. Wide zone of the media population inversion behind the nozzle unit (10-15 cm) stipulates simplicity of selection of the scheme of the resonator for obtaining laser generation with maximum efficiency. Thus the characteristic size of a mirror of the resonator is approximately equal to length of a zone of generation, i.e. 10-15 cm, therefore manufacturing of these mirrors does not call serious technological difficulties.

The capability of GDL activity at high pressure of gas in the resonator in front of the nozzle unit (up to 3 MPa and more) provides with a straight line an exhaust of spent gas in atmosphere with the help of the supersonic diffuser of a special design. Such capability all remaining considered lasers are deprived. The aft ejector is necessary for maintenance of a direct exhaust of these lasers which requires additional gas flow rate, that complicates the installation and reduces specific energy of radiation.

It is necessary to note also, that GDL output power rather weak varies at change over a wide range of parameters of gas (pressure, temperature, chemistry) in the combustion chamber, i.e. GDL is not critical to accuracy of a task and maintenance of an operational mode. One more virtue GDL is the small time (some seconds), necessary for start.

Thus, GDL is the rather simple, reliable and flexible tool ensuring high output power of laser radiation.

### 2.2.2. Mobility

From all considered types of lasers GDL for today has, apparently, least "weight" and "volume" of 1 kW of laser power. It allows rather simple to place GDL on any vehicle, whether it will be airplane or helicopter, ship or railway platform, etc. Such GDL mobility allows with its help to solve many tasks, including connected with disposal of petroleum films.

### 2.2.3. Selection of propellant components

One of specific singularities of working process of the $CO_2$-laser is, that the working mixture of gases should contain in a receiver rather small quantity of water vapor (no more than 5-6 volume %). On the other hand, the operation temperature of gas in a receiver should be enough high (1300-1600 K). It superimposes limitations on selection of propellant components (fuel and oxidizer). From the point of view of maximum specific energy of radiation (J/kg) optimum components are gaseous at standard conditions with damp (CO) and liquid nitrous oxide ($N_2O$), thus additional nitrogen (or air), necessary for working process, moves in a receiver from a separate source, and the water vapor in quantity 1,5-2 % is formed by burning small quantity of hydrogen or alcohol in air.

### 2.2.4. Wavelength of radiation

The analysis of the data presented in tab. 1 show, that for the solution of a problem of disposal of petroleum films, the optimal wavelength of radiation is that of the $CO_2$-laser, as this radiation is weakly absorbed by the film and is strongly absorbed by water. From the point of view of a diffraction limited angular divergence of radiation, the $CO_2$-lasers (EDL and GDL) lose for CL. However with reference to a considered problem, the value of a divergence is not critically important. The estimations show, that at length of a beam from the laser source down to a surface of water ~ 50-100 m and power of radiation up to 250 kW, the precision focussing of laser beam is not required, for the reason of obtaining required density of power in a spot on water surface. It is possible to utilize unfocussed or partially focussed beam. Therefore with reference to a considered problem, the relatively large wavelength of GDL radiation is not the factor of insufficiency.

The explained above reasons allow to make a unequivocal conclusion for the selection GDL as the laser radiation generator for the mobile installation intended for disposal of a petroleum film on a surface of water. At the estimated power of radiation, GDL is preferable in weight and volume factors in comparison with EDL, HCL and COIL: it does not require the electric power, use to utilize a low toxic fuels, provides the direct exhaust of utilized gas to atmosphere. GDL is very simple in control, it is not critical to changes of working parameters in a sufficiently broad range, and it is convenient in operation. The important circumstance is that we had in our hands a very reliably and operable GDL with output power 100 kW, which in our days is a very effective tool for intensive research program to be carried out. In particular, the design of the aerodynamic window permitting to extract a laser beam from the resonator zone to atmosphere without application of transparent for the working wavelength materials has been effectively developed.

## 3. Selection and substantiation of basis performances of the laser installation and GDL principal diagram

In the previous chapter the comparison of various types of high power lasers was carried out from the point of view of their application for the solution of this task, where was shown, that the most appropriate type of the laser is $CO_2$ - GDL. However, there are different types of GDL therefore it is necessary to choose among them the optimal version, and also to develop the general concept of the power installation for optimal supply of working components in GDL.

In the present chapter the analysis of the various GDL schemes, their advantages and disadvantages are given from the point of view of application in the considered installation, the selection of propellant components for GDL is justified from the point of view of their power efficiency, production and toxic characteristics, and ecological safety.

Last part of the chapter is devoted to the substantiation of the GDL characteristics working on the chosen components. For this purpose the calculation results which have been carried out according to [2], are permitting to define specific power of laser radiation at the given initial parameters, such as fuel chemistry, temperature and pressure in front of the nozzle block, expansion ratio of the nozzle etc are used. The calculation results and computer analysis are shown.

For definition of the main dimensional characteristics of the laser and its flow organizing parts it is necessary to define the basic level of laser radiation power, which, in turn determines the total fuel consumption and remaining geometrical characteristics of the laser.

## 3.1. Selection of optimal radiation power of GDL

### 3.1.1. Theoretical substantiation of power level

The basic capability of disposal of petroleum contaminations from water surface by means of irradiation of a petroleum film by a high flux of laser radiation has been proved experimentally and presented in our publications. However, the published data give only the general representation about basic possibility of using of such method. It is obvious, that the combustion is complex multiparameter process and, if we talk about operational use of the mentioned principles in the particular technical device intended for practical use, the optimization of parameters of the installation is necessary. The important value in this case is the efficiency, which it is possible to understand as the area of cleaned water surface divided to mean energy, used on clearing, ($m_2/kW$). For optimization of parameters of the installation it is necessary to conduct a cycle of research works on study of evaporation, ignition and combustion processes of petroleum used for the above mentioned method. The experimental research of the given problem even without qualitative understanding of involving processes represents a very difficult task in connection with vast quantity of optimization parameters and requires heavy material costs and time. In this connection the preliminary development of theoretical model representing process of disposal of petroleum film is desirable.

The second phase [3] of the process is presented, the various interaction mechanisms of a laser beam with a petroleum film on a water surface are theoretically considered. The following mechanisms are included:

-   laser heating active absorbing, translucent and transparent films;
-   an explosive boiling-up and conditions of process efficiency;
-   evaporation and vaporization.

The basic conclusions of the technological approach are the following.

1.  For reaching temperatures of evaporation of the film the necessary time of radiation action is about 100 nanoseconds at laser power 100 kW and diameter of an irradiated zone is about 30 cm. Temperature of water layer under the surface of film can essentially exceed boiling point. Such situation arises due to superior velocities of laser heating (up to several millions degrees of Celsius per one second) and inertial character of heat transfer process. At such temperatures the water passes in a metastable state with active outgassing. Under a surface of a film there will be micro explosions of air - steam bubbles. As result, there will be a separation of a film from the surface of water, after it`s breaking the outflow of drops of petroleum take place to the open air.
2.  The scanning velocity at power of the laser of 100 kW and diameter of a spot 30 cm ought to be in range of 10-12 m/s. Thus the velocity of cleaning of a surface 18000-24000 m$^2$/hour can be provided.

It is necessary to note, that all calculations are fulfilled for thickness of a petroleum film about 100 microns, which is rather large value and is observed only in the initial moments of formation of oil spillages. For much thinner films, the velocity of cleaning will be increased proportionally.

### 3.1.2. Experimental researches

For experimental confirmation of basic physical principles included in the basis of the Project, the participants carried out preliminary experimental research of interaction of laser radiation with films of various petroleum types on a surface of water. The research works were conducted within the framework of financing of activities.

During our experimental research works the GDL developed by our team, and prepared for this particular task to be solved had been used. The optical scheme of realization of experiments is presented in Fig. 1.

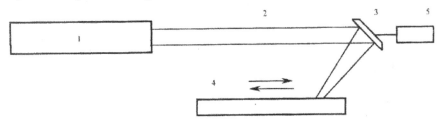

**Figure 1.** Optical scheme of oil film interaction with laser beam.

The GDL 1 generates radiation 2 with a wavelength 10,6 microns. The radiation is directed on a concave focusing mirror 3 and, being reflected, impacts on a surface of test heterogeneous structure in a cuveta 4. The mirror realizes periodic oscillations with the help of the mechanical device 5, thus the laser beam scans on a surface of a dish. Diameter of laser beam cross section on a surface of a dish changed from 15 cm to 30 cm. The laser radiation power was within the limits of 100 kW.

In Figs. 2, 3 and 4 the results of experimental research works are presented, on which it is possible to make following preliminary conclusions.

**Figure 2.** Intensive combustion of petroleum contamination under laser radiation.

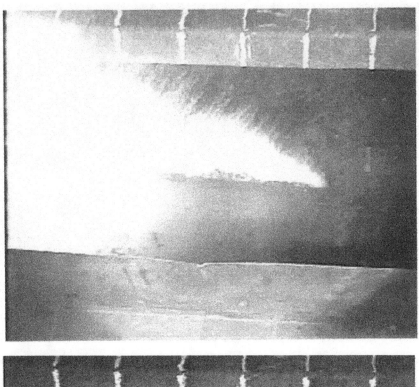

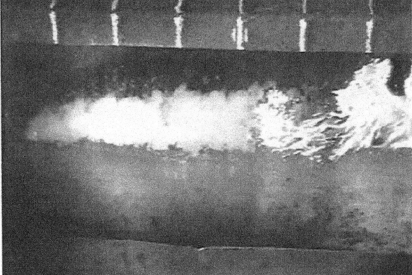

**Figure 3.** Combustion of kerosene under laser radiation.

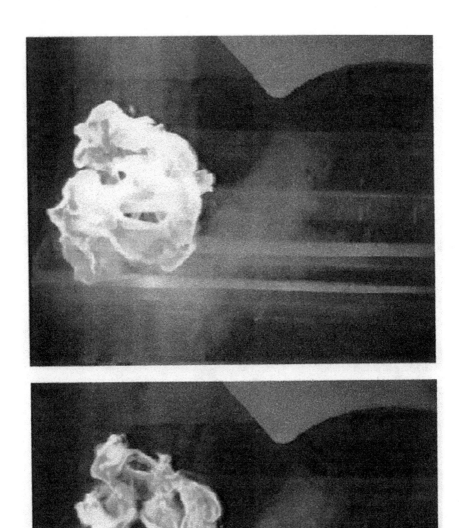

**Figure 4.** Combustion of engine oil under laser radiation.

1. The carried out experiments testify the possibility of using of the laser installations for cleaning of an aqueous medium from contaminations of petroleum and of it processing products.

2. For definition of optimal parameters of the laser installation and operation modes in practical conditions it is necessary to select experimental conditions of the working regime according to the special program approved by a potential customer.

3. For necessary reserve, the radiation power of GDL of 100 kW for the installation to be developed according to the present consideration is chosen.

## 3.2. Selection of the GDL scheme and its substantiation

### 3.2.1. Conventional scheme of GDL

The principle of GDL operation [4] is based on non-equilibrium fast expansion of the heated up working gas in the supersonic nozzle, at which there is a partial freezing population of oscillatory levels of molecules $CO_2$ and $N_2$ and formation of population inversion of upper and low laser levels. The heating of gas to an operation temperature 1300 –1700K at pressure 2-3 MPa is conducted in the combustion chamber through of combustion of chosen fuel – oxidizer components. The transformation of the accumulated oscillatory energy of gas to laser radiation has taken place in the optical resonator located down the nozzle. After the resonator the working gas passes through the supersonic diffuser and will be exhausted in atmosphere.

The most fast gas expansion take place near the critical zone of the nozzle at the length equal to the value of one order of magnitude of the critical cross-section size. Therefore for fast freezing of oscillatory energy it is necessary to use the nozzles with a small size of critical cross-section (< 1 mm). On the other hand, the pressure of gas after the nozzle (i.e. in the resonator region) should be small enough (< 100 mm Hg) to reduce to a minimum the losses of oscillatory energy due to collision relaxation. It has been solved by use of nozzles with large expansion ratio (area ratio of output to critical cross-sections $A / A^* \geq 20$, that corresponds to the value of a Mach number M 4).

For effective transformation of the accumulated oscillatory energy to laser radiation the amplification of a small signal during the pass the length of an active zone should be large enough to compensate optical losses (removal of radiation, absorption in mirrors, dissipation in medium) and to ensure a high enough radiation power in the resonator. In GDL of very low powers (tens W) is possible to use the so-called mononozzle, which represents the flat (rectangular) nozzle with critical cross-section as a narrow slot (Fig. 1,a). However for high-power GDL implementation of the mononozzle is practically impossible because of its large length (1 m and more); arising thermal deformations result in essential to nonuniform width of critical cross-section, h *, down to full blocking of critical cross-section in some zones. Besides the mononozzle does not allow to receive enough large transversal size of an active zone, that causes certain problems at resonator selection. Therefore in high power GDL the so-called nozzle unit representing set of a many relatively short (~100 mm) slot-type nozzles (Fig. 1.1, a) is used. These nozzles were formed by contoured nozzle blades, made from thermal resisting metal. Such design of the nozzle unit allows changing in a wide range width and height of an active zone. Besides the nozzle unit provides a uniform enough density field of gas in an active zone, that allows realizing a small angular divergence of GDL radiation.

In view of such advantages of nozzle blocks, this technology now is widely used for GDL, and therefore it is possible now to name such GDL systems as the "GDL of traditional scheme". The radiation power of such a GDL is in our days up to 250 kW and more for continuous and P-P radiation modes of operation [5, 6].

### 3.2.2. GDL with a set of axisymmetrical nozzles

The nozzle unit of the conventional scheme has the disadvantage that it is complex in manufacturing, and also is subject to influence of thermal deformations. There is an alternative approach to the development of nozzle units, which consists of the following. Instead of two-dimensional nozzles the axisymmetrical nozzles with small diameter of critical cross-section are used. The supersonic part of such nozzle can be made as a conical or contoured. In this case nozzle unit looks like a plate, where the numerous small-sized nozzles located close as possible to each other (Fig. 1.1, c) are made. The manufacturing of such a nozzles is possible, they are not subject of thermal deformations, by the appropriate arrangement of these nozzles on a holding plate (quantity of rows and quantity of nozzles in a row) it is possible to give output cross-section of the nozzle unit with required configuration.

However such nozzle unit has at least two disadvantages caused by presence of "empty" zones between nozzles, since complete filling of output cross-section by round nozzles basically is impossible.

The first disadvantage consists of the empty zones, which are the sources of shock waves in a supersonic flow. That wave results in density inhomogeneous structure of gas in a zone of the resonator and, hence, leads to increase of an angular divergence of laser radiation.

The second disadvantage consists of the effect, which creates in empty zones the viscous vortices flows, which results in reduction of total pressure of a main flow. In turn, the reduction of total pressure has negative consequences on the supersonic diffuser operation, which can be appreciated as reduction a pressure recovery coefficient. In this case for realization of direct exhaust of working gas to atmosphere it is necessary to raise combustion-chamber pressure, which results in the decrease of output power of laser radiation.

For these two reasons GDL with the nozzle unit as a set of axisymmetrical nozzles is not widely used at the noticeable production scale.

### 3.2.3. GDL with mixing of working gases in a supersonic flow

In the case of traditional GDL the simultaneous heating of all components of a working mixture - $N_2$, $CO_2$, $H_2O$ has taken place in the combustion chamber. However initial reserve of laser energy contains, primarily in oscillatory - exited molecules $N_2$, which part in gas mixture makes ~ 90 %. The role $CO_2$ and $H_2O$ consists only of transforming accumulated in $N_2$ oscillatory energy to laser radiation.

The pure nitrogen has extremely large time of an oscillatory relaxation stipulated by collisions of molecules $N_2$ among themselves. As the result, the oscillatory levels $N_2$ are easily frozen at fast expansion of the working mixture in the nozzle. However other components of a mixture ($CO_2$ and $H_2O$) considerably accelerate a collision relaxation $N_2$ and by this reduce a reserve of oscillatory energy.

For increase of a reserve of oscillatory energy it is desirable to raise temperature in the combustion chamber. However in the traditional GDL temperature is limited by value about 2300K; the dissociation $CO_2$ begins at higher temperatures, and the efficiency of conversion of the accumulated oscillatory energy in laser radiation decreases. In contrast to $CO_2$ the molecular dissociation $N_2$ begins at temperature ~ 4000 K.

The presented ideas have resulted in the new GDL scheme, which concept consists in following: the pure nitrogen, heated up in a prechamber, expands in the supersonic nozzle, and at the exit of the nozzle to it are mixed cold $CO_2$ and $H_2O$. Such scheme has three basic advantages for the usual scheme GDL: a) temperature of nitrogen in a prechamber can be essentially increased, b) more effective freezing of oscillatory energy of nitrogen is reached, c) energy loss for $CO_2$ and $H_2O$ heating is not necessary.

In view of the indicated advantages mixing GDL can basically ensure the increase of specific power of radiation many times in comparison with the GDL conventional scheme. A main problem arising at realization mixing GDL - necessity to ensure fast intermixing hot $N_2$ with cold $CO_2$, but the disturbances of a supersonic flow $N_2$ should be kept at minimum.

Some solutions of this problem (Fig. 1.2) are possible. Technically it is much simply to organize mixing of $CO_2$ and $N_2$ in wakes (Fig. 1.2, a). In this case flow $CO_2$ moves in parallel to flow $N_2$, in this case two sets of supersonic nozzles - one set for N2, second for $CO_2$ are used. A advantage of such scheme of supply is that arising at interaction of wakes disturbances are minimum in comparison with other schemes of intermixing.

The basic disadvantage of this scheme is that the mixing of wakes is rather slowly, therefore for effective work of the laser the working section (from nozzle exit down to the diffuser inlet) should have sufficient length, which results in large relaxation losses and reduction of a pressure recovery coefficient of the diffuser. For these reasons the scheme GDL with intermixing of supersonic wakes, apparently, is not effective.

The theoretical analysis and the experimental data show, that the GDL schemes are more effective, in which the injection of cold $CO_2$ is made near to critical cross-section of the nozzle in it transonic part (Fig. 1.2,b,c). In experiments with heating $N_2$ in electric arc plasma generating device and supply of $CO_2$ under the scheme 1.2,c, was received record for GDL value of a gain equal 3 % cm⁻¹, that exceeds maximum value of a gain for usual GDL almost to an order of magnitude. However specific power of radiation was insignificant, that is connected to a small scale of the installation.

Despite of basic promising character of mixing GDL, there is a lot of problems connected with their practical realization.

One of the most important problems is obtaining hot nitrogen at temperature 3000-4000K. Application of electric arc plasma generating device for high-power mixing GDL is unreal, since at the consumption of nitrogen 10 kg\sec and more required electrical power exceeds 100 MW. It is possible to use for obtaining hot nitrogen special fuels, however all of them are strong toxicants.

As a source of hot nitrogen, the devices can also be considered, in which a combustion of metal in air has taken place. As a result of this strong exothermal reaction hot nitrogen and oxide of metal as particles are formed, which are extracted from the formed flow with the help of a cyclone separator. However, necessity of a very high degree of a nitrogen stream cleaning from particles makes this way problematic.

Other important technical problem is the development of the nozzle (selection of a material, development of a robust design, which would maintain without destruction large heat flows in the region of critical cross-section.

The supersonic diffuser causes one more important problem. For an effective work of the laser translation temperature of working gas in an active zone should be sufficiently low, whereas temperature in a prechamber should be relatively high. For the coordination of these two conditions it is necessary to use supersonic nozzle with large Mach numbers (M=5 and more). However is known, the higher Mach number is, the lower is pressure recovery coefficient for the diffuser. Therefore for a direct exhaust of gas in atmosphere it is necessary to develop special diffusers, thus can appear, that such task basically is impracticable without use of a fodder ejector. However consumption of inducing gas in some times exceeds the consumption of induced gas, and in this case specific laser power designed on summarized gas flow rate, considerably decreases.

From explained above it is possible to make a unequivocal conclusion: taking into account problems connected to development of the high power mixing laser, and also absence of the prototypes of such lasers, for the solution of a task of disposal of petroleum film on water surface it is necessary to choose GDL of conventional scheme.

### 3.3. Selection of fuel for mobile GDL

In the present part the analysis of possibility of using various fuels for support of operation of the mobile $CO_2$-GDL, intended for disposal of petroleum film on a surface of water is carried out.

$CO_2$-GDL provides the laser radiation generation in a continuous/P-P modes of operation with radiation wavelength 10.6 microns. A basic physical principle of operation gas dynamic of the $CO_2$-laser is the fast expansion in the supersonic nozzle of a mixture of gases ($CO_2$, $N_2$, $H_2O$) preheated up to temperature 1500K- 2000K. In a supersonic flow with Mach number M = 4 - 5 temperature of gases down to values 300K – 350K, necessary for population inversion of laser levels obtaining of a molecule $CO_2$. For nozzle operation at direct exhaust in atmosphere after pressure recovery in the diffuser (the static pressure in a zone of generation is equal 5 $10^3$-1 $10^4$ Pa) the nozzle inlet pressure (stagnation pressure of a flow), equal 2-3 MPa is necessary. At small transversal and longitudinal sizes of the supersonic nozzle (the height of critical cross-section is equal 0.3 - 0.5 mm, and length of the nozzle 20 - 40 mm) characteristic time of the gas flowing through the nozzle appears small and comparable with time of oscillating relaxation of molecules of a mixture of gases, i.e. oscillating - oscillating exchange (*V-V* process) and oscillating - translation relaxation (*V-T* process) has taken place in non-equilibrium mode. As a result of non-equilibrium processes in oscillating exited molecules at certain gas mixture ratio and flow parameters in the supersonic nozzle the inversion population of the upper laser level of molecules $CO_2$ (0001) is formed. The molecules $CO_2$, are light generating molecules (laser radiation is realized on transition $00^01$ -$10^00$), $N_2$ molecules are donors molecules transmitting oscillating energy to the upper laser level of molecules $CO_2$ ($00^01$), the molecules $H_2O$ are intended for population of the lower laser level of molecules $CO_2$, ($10^00$). The optimal ratio of mixture of gases for an effective work of gas dynamic $CO_2$-laser should contain $CO_2$ = 0.1, $N_2$ = 0.89, $H_2O$ = 0.01 (volumetric fraction). It is necessary to note, that the increase of a volumetric vapor fraction of ($H_2O$) results in decrease of the laser radiation power with other things being equal.

For mobile GDL, laser radiation, ensuring high power in CW/P-P modes, a basic method of thermal excitation is the combustion of such fuel components, which as final products give necessary components of a laser mixture at relatively high temperature. The optimal mixture ratio of gases at high temperature of mixture can be obtained, for example, at burning of carbon oxide (CO) and hydrogen ($H_2$) in air or other oxidizer ($N_2O$, $N_2O_4$) with subsequent use of ballast nitrogen.

The specific requirements to the analyzed mobile installation impose some limitations on use this or that fuel, oxidizer and ballast gas. The main difference of this installation from existing lasers is the requirement of large duration of operation.

For processing of the greatest possible area of the polluted basin, the laser installation should have an operating time appropriate to operation capabilities of a vehicle, on which it is secured. For the helicopter this time makes $t \sim 30$ min. The carried estimations of interaction of a beam with a water surface covered with thin petroleum film, have shown, that the power losses by heating water can be reduces through fast scanning of a surface by a beam of power $W \sim 100$ kW. These requirements allow estimating total of fuel and ballasting gas necessary for mobile GDL operation. The specific power ($w_{out}$) existing homogeneous $CO_2$-GDL, as a rule, does not exceed 10 kJ/kg. Thus weight of fuel and ballast gas for such GDL operation should be not less than $m = Wt / w_{out} = 18000$ kg. This value essentially will increase at the account of weight of the equipment for storage of working components. Apparently, that such large weight is unacceptable for laser arrangement on a flight vehicle.

At the same time, the main part (80 95 %) of weight of design components used for creation of a propulsive mass GDL, is the weight of an oxidizer and ballast gas. Thus, use of free air as an oxidizer and a ballast gas is a unique opportunity to guaranty this installation operation. As $CO_2$-GDL fuel usually carbon monoxide is used with the hydrogen addition.

Carbon monoxide allows receiving high values of specific power of radiation in comparison with liquid fuels, such as toluene, benzole, kerosene, but weight this fuel together with a system of storage is rather high. Weight CO is 13 - 15 % from total mass of spent components, i.e. in our case of 2000 - 3000 kg. Weight of a storage system of such quantity gaseous carbon monoxide is $\approx 7000$ kg, and summarized weight fuel and system of storage $m_\Sigma$ = 9000 - 10000 kg. In case of use cryogenic state carbon monoxide summarized weight $m_\Sigma$ can be reduced approximately twofold, but nevertheless it considerably surpasses appropriate weight for liquid hydrocarbon. Weight of liquid hydrocarbon is 4 5 % from total mass of spent components, and the tanks for storage are much lighter than for carbon monoxide, as the hydrocarbon are stored at atmospheric pressure. In our case weight of liquid hydrocarbon is 700 - 900 kg, and summarized weight fuel and system of storage $m_\Sigma$ = 1000 - 1500 kg.

Taking into account the mentioned above reasons, in the present operation as possible fuel are considered benzole, toluene and kerosene, and as an oxidizer and a ballast gas (instead of nitrogen) free air compressed by the compressor.

As it was mentioned above, for normal operation of supersonic nozzles of the laser it is necessary to ensure high total pressure of a stream ($P_0$ = 2 - 3 MPa). The aviation compressors can ensure a compression ratio up to $P_0/P_a$ = 20. Here $P_0$ stagnation pressure of a working mixture in a GDL channel, $P_a$ atmospheric pressure. In all further calculations the value of stagnation pressure of a flow $P_0$ = 2 MPa and maximum values, appropriate Mach number of the nozzle M = 4.7 with the area ratio of an exit of the nozzle to the critical cross-section $F/F^* \approx 25$ was accepted.

At comparison various fuels, basic parameter is the power efficiency, namely specific power of laser radiation $w_{out}$ (power of laser radiation referred to summarized mass flow of laser components), achievable at use of the given fuel. In this connection the analytical-theoretical research of influence of various parameters on value $w_{out}$ was carried out.

Temperature and chemistry of a working mixture formed at combustion various fuels in air, were determined by thermodynamic calculation. The necessary for calculation enthalpy of initial components was taken from [4]. For the reason of uncertainty of efficiency of the compressor (approximately it is 0.75-0.83) in a main part of executed calculations temperature of inlet air in the combustion chamber was assumed to equal appropriate temperature at adiabatic compression up to value

$P_0 / P_a$ = 20. This temperature calculated according to the following formula

$$T = T_a \cdot \left( \frac{P_0}{P_a} \right)^{(\gamma-1)/\gamma} ,$$

where $T_a$ - temperature of free air, $\gamma$ - isentropic exponent. At adopted values $T_a$ = 298K, $\gamma$= 1.4 temperatures of inlet air in the combustion chamber is $T$ ~700K. The enthalpy, appropriate to this temperature of air is equal $I$ = 400 kJ/kg.

At thermodynamic calculation the thermal losses which depend on a design of the combustion chamber and beforehand are not known were not taken into account. In this connection in the subsequent calculations of flow of a working mixture in the nozzle block inlet temperature $(K)$ was taken equal to 0.95 of temperature obtained in thermodynamic calculation.

The thermodynamic calculations have shown, that the working mixture contains in the investigated temperature range (or oxidizer-to-fuel ratios) only necessary for operation $CO_2$ - GDL components ($CO_2$, $N_2$, $H_2O$) and oxygen ($O_2$), as the content of remaining components is insignificant. Therefore in the further researches it was supposed, that the working mixture consists only of carbon dioxide, nitrogen, water and oxygen.In the paper the calculations of non-equilibrium flow of a mixture $CO_2$ - $N_2$ - $H_2O$ - $O_2$ in the supersonic nozzle are carried out as well. The kinetic model describing population change of separate oscillating levels of molecules $CO_2$ and $N_2$ at processes of VT- and VV- of exchange was used. All oscillating levels of a molecule $CO_2$ with characteristic temperatures $\theta < 5000K$ were taken into account. It was supposed also, that the velocity of exchange of oscillating energy to a Fermi-resonance considerably surpasses a velocity of all remaining VT- and VV-processes. Thereof, the modeling system of levels was introduced, some of which have a large degree of degeneration and correspond to several levels in a full system of levels of a molecule $CO_2$. As a result of calculations the population inversion and gain coefficient of the nozzle unit outlet was determined. The calculations of specific output power of laser radiation were conducted at following input data.

1. The working mixture of gas dynamic laser is formed at combustion hydrocarbon fuel in air. The stagnation pressure of a flow is equal 2 MPa, total temperature is equal 1600K.
2. The nozzle unit of the installation consists of flat contoured nozzles with the height of critical cross-section 0.8 mm and expansion ratio 25.
3. The sizes of exit cross-section of the nozzle unit: length - 1600 mm, height - 130 mm.
4. The size of a cavity of the resonator along the flow direction - 200 mm.
5. The three pass unstable resonator with a multiplication factor

$M = 1.4$. A reflection coefficient of mirrors $r = 0.98$, the losses on dissipation make $\circledcirc = 0.03$ m$^{-1}$.

The contour of a supersonic part of the nozzle was obtained as follows. The flow of gas with an isentropic exponent equal 1.35 in the contoured nozzle with an angular point was calculated and as a contour of a supersonic parts of the nozzle the form of a streamline $\circledcirc = 0.9$ was used.

| Fuel | $k_\circledcirc$, m$^{-1}$ | $w_{max}$, kJ/kg | $w_{out}$, kJ/kg |
|---|---|---|---|
| Benzole | 0.41 | 23.6 | 9.6 |
| Toluene | 0.39 | 23.1 | 9.2 |
| Kerosene | 0.32 | 20.8 | 7.4 |

**Table 3.** Values of specific output power of laser, based on different fuels.

Thus, value of specific output power of laser radiation, which it is possible to expect at use as fuel benzole and toluene, and as an oxidizer and ballast gas - dry air, is $\approx$ 9 9.5 kJ/kg. are presented in Table 3. At use of kerosene this value is approximately 2 kJ/ kg less.

By selection of fuel, except for power efficiency, it is necessary to take into account production and toxic characteristics, and cost of components.

The application of benzole is inconvenient at low environment temperature. The freezing temperature is 5.5°C. For toluene this value is equal -80°C, and for kerosene -38°C.

The limit for a room vapor concentration of researched fuel has following values: 5 mg/m³ for benzole, 50 mg/m³ for toluene, 300 mg/m³ for kerosene.

From said it is possible to conclude, that the application of benzole is unwise, as it a little surpasses toluene according to power efficiency, but is significantly worse for toxic than toluene and is less convenient in operation.

As to selection between toluene and kerosene, the application of toluene provides specific power increase approximately 25 %. However use of kerosene has a number of advantages. At first, kerosene is approximately 3 times cheaper than toluene. Secondly, kerosene is fuel for airplane and engines. Therefore at the arrangement of the laser installation working on kerosene on a flight vehicle the additional system for fuel storage is not required.

As conclusion for this chapter it is possible to note the following.

The homogeneous laser working on combustion hydrocarbon fuel products is most reasonable for development of the mobile installation.

The specific requirements to the mobile laser installation,impose the limitations on use this or that fuel, oxidizer and ballast gas. The main limiting factor is the required long duration of continuous operation. The carried out estimations have shown, that summarized weight of fuel, oxidizer and ballast gas necessary for GDL operation, is very hight and is unacceptable for arrangement on a flight vehicle. At the same time, the main part of weight of fuel components used for GDL operation mass, is weight of an oxidizer and a ballast gas. In this connection, use of free air compressed by the aviation compressor, as an oxidizer and a ballast gas is the only way to organized operation of this installation.

The carried out comparative analysis of various fuel has shown, that most reasonable among them are toluene and kerosene. The application of toluene provides specific power of the installation approximately 25 % higher than that of kerosene. However, use of kerosene has a number of the advantages. At first, kerosene is approximately 3 times cheaper than toluene. Secondly, kerosene is the fuel for airplane and helicopter. Therefore at arrangement of the kerosene operating installation, on a flight vehicle the additional system for storage fuel is not required.

# 4. Selection of the power installation for GDL

## 4.1. Selection of a schematic of the power installation

### 4.1.1. Special conditions

In the previous chapter devoted to the computational substantiation of the basic power characteristics, the capability of development of the mobile $CO_2$ - GDL with use of kerosene as fuel, and as an oxidizer and a ballast gas - atmospheric air supplied by the compressor is shown.

Besides, some GDL parameters, permitting to determine the required characteristics of the compressor are added.

The calculations have shown, that in case of GDL the condition Po /Pn = 20 should be realized

Where:    Po    -    stagnation pressure of a working mix(mixture) in GDL channel;
            Pn    -    atmospheric pressure.

In view of channel losses from the compressor to GDL inlet of the chosen compressor should have the pressure ratio not less than 22 - 23.

Temperature of inlet air in GDL combustion chamber chosen at calculations is 720K, which corresponds to the temperature range of compressed air for compressors with Pк > 22÷23.

For an estimation of necessary value of airflow rate through GDL it is rational to set specific energy extraction at known accepted power range of 100 kW. As was shown in the previous chapter, the GDL specific power with combustion of kerosene in air is 7,5 kJ/kg. However this value can be reached at dry air only. At increase of humidity of atmospheric air the specific power begins to decrease and at humidity value 100 % (T = 25ºC) it is 5,4 kJ/kg.

To achieve the higher specific power it is possible in principle to include air dryer in an air flow path, however it large overall dimensions and power consumption can result in technical and operational problems and, besides the automatic control of dryer operation is necessary if air humidity and fuel have changed. It will be more conveniently to have certain losses of specific power at humid air, having compensated losses by increased air flow.

Therefore, having accepted average specific power of 6 kJ/kg, instead of 7,5 kJ/kg we shall receive an estimated value of an air flow - 16,25 kg / s.

Thus, for a operation of the laser installation in power range of 100 kW it is necessary to have the compressor with an air flow rate ≈ 16 kg / s and Pк >22÷23.

In essence task of air compression at flow rate of 16,25 kg /s and pressure in access of 20 atm does not contain unsolved technological problems, since compressor engineering is widely applied in industry. The task consists to find the optimal solution ensuring the reasonable cost of the installation, reliability of its activity and convenience in operation. Let's consider three possible schematics of power installation.

*4.1.2. Development special power installation for the above mentioned parameters of the compressor*

In this case the installation will consist (Fig.5) of the compressor itself, turbine for driving, gas generator for rotation the turbine, and also fuel supply system, control and regulation units.

As there is no installation for the mentioned parameters in industry, it is necessary to develop such installation from very beginning, to conduct preproduction activity and to find a producer. For all reason, the price of the produced installation will be high, and development - time consuming.

Besides, short order book for such installations (single production) will lead initially to unprofitable operation for any producer.

### 4.1.3. The installation of air compressor on aviation engine shaft

In general approach, the task can be decided rather simple. On the shaft of the aviation gas-turbine engine the compressor with the required characteristics is installed instead of the propeller.

The calculations show, that the shaft horsepower about 5,5-8 MW is necessary for compressor driving with the flow rate of 16 kg / s and Pк 23 kg / cm². It is power of the aviation gas-turbine engine of middle power class, a lot of which is in series production. However, as engineering realization the task has a number of serious problems.

The nominal speed of the propeller shaft of existing aviation engines is in the range of 5-8 thousand rpm, and all high efficiency centrifugal compressors require operation speed of the order above. For example, industry compressor ТВ7-117 being far from the best in efficiency already requires 30000 rpm.

The installation between the gas-turbine engine and compressor of the high-speed gear reducer (multiplicator) because of complexity in operation and large overall dimensions excludes such solution as acceptable.

Technically task is solved by the installation on the shaft of the engine of an axial multistage compressor (Fig.6). But even the principal diagram general view demonstrates the technical irrationality of such solution, it practically two aero-engines, one of which a production one, and the second requires large design and manufacturing costs connected refurbishment for realization of rigidity of the shaft, organization of turbine side exhaust etc. Besides of that, the overall dimensions and accordingly the weight of the installation are considerably increased.

### 4.1.4. Air bleed from aero-engine compressor final stage

As it is visible from the scheme Fig.7, if it is decided to take an air bleed for GDL after a final stage of the engine compressor, practically there is no necessity redesign the power installation with installing new aggregates. Moreover, the dismantling from the engine of the free running turbine operating for driving the propeller shaft for the reason of propeller nonuse, considerably simplifies a design of the installation as a whole.

This solution bribes by the simplicity, since the engines not only will be used with minimum modification, but also all control systems are kept, including the use of the control panel which is taken off from a flight vehicle.

The new developing installation

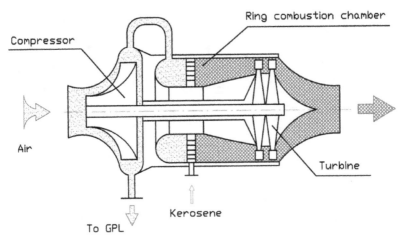

**Figure 5.**

Besides, from point of view of the power, dismantling of the free running turbine together with nozzle block allows to lower required power of the turbine of system of the compressor - free running turbine.

The decrease of required power of the system offers the possibility to reduce quantity of required air and kerosene in the combustion chamber of the gas-turbine engine. With constant power rate of the turbine for the drive of the compressor and of the compressor flow rate, probably, it exist the possibility to take off a part of the air and to direct it to the combustion chamber of gas dynamic laser. At the same time, utilizing the characteristics of the combustion products down the main turbine ($\rho$, m, T), it is possible to organize an ejection of GDL gases down the supersonic diffuser for improvement of the gas dynamic laser overall performance, especially at start.

## 4.2. Selection of the aviation gas-turbine engine

With the purpose of optimal selection of an aviation-engine for this task it is necessary to formulate the initial requirements.

1.  For obtaining air pressure in GDL inlet not less than 20 atm and in view of losses in a flow channel the compressor of the engine P$_K$ should be in the range of 22-25, not less.
2.  The GDL inlet air flow rate is 16,25 kg /s.

As for elimination of extreme engine power reduction the air bleeding from the compressor should not exceed 15-18 %, the total air flow rate through the engine should be not less than 100 kg /s.

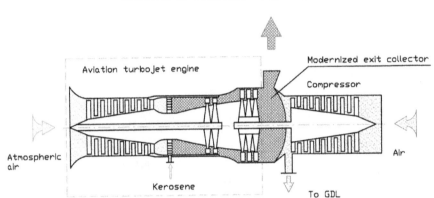

**Figure 6.**

3.  The chosen aviation engine should widely be used by standard aviation operation, that enables to utilize engines which are decommissioned from flight vehicles, having passed prescribed operational life limit, but still are operable for ground installation. To buy them is much cheaper, than to buy new engines.
4.  It is desirable to make a choice of domestic (Russian made) engines that enables in case of realization of the project to connect to activities, both engine developing company and manufacturer.

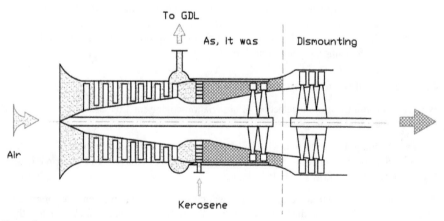

**Figure 7.**

The table data of some Russian engines characteristics: Пк and G of airflow rate (kg / s) are given below.

| Engine | HK-12CT | HK-14CT | HK-14Э | HK-16CT | HK-17 | HK-36CT | ПС-90A | РД-33 |
|--------|---------|---------|--------|---------|-------|---------|--------|-------|
| Пк | 8,8 | 9,5 | 9,5 | 9,68 | 9,68 | 23,12 | 19,6 | 22,0 |
| G (kg/s) | 56,0 | 37 .1 | 39,0 | 102,0 | 102,0 | 101,4 | 56,0 | 80,0 |

**Table 4.**

The analysis of the table results in a conclusion, that most reasonable for the laser installation could become the engine HK-36CT serially produced by Samara plant «Motorostroitel», which satisfies to all requirements presented above. It is obvious, this is not the single engine, which can be put to use for this task, but in advantage of it would be desirable to note the following.

The high performance engine HK-36CT of the Samara Technological complex «Motorostroitel»," is developed in 1990 on the basis of an aero-engine HK-321 and is designed for the drive of the centrifugal supercharger in a structure gas pipeline pumping aggregate.The modular design of the engine facilitates transport and assembly. The engine has the remote control panel and record of long life ground operation. The general view of the engine is shown in Fig.8. In Fig.9 the design concept is shown.

**Figure 8.** Gas-turbine engine NK-368T

### 4.3. Gas dynamic assessment of the chosen aero-engine as the power unit for mobile GDL

In the previous chapter the task of GDL supply with oxidizer (air) by use of the HK-36CT engine compressor was considered.

However, one more function - use of the engine exhaust gas as working medium for ejection of a GDL exhaust system is planed to the chosen engine.

The practice of development GDL with direct exhaust of combustion products in atmosphere has shown, that the aft diffuser of the laser produced the pressure recovery of gases up to atmospheric pressure requires for start significant the combustion chamber pressure increase (more than 30 atm), or decrease of exit pressure below atmospheric one. It is possible to ensure the last requirement, through installation of an ejector, for which the exhaust gases down the engine turbine are the working medium.

Design schematic of NK-36ST engine

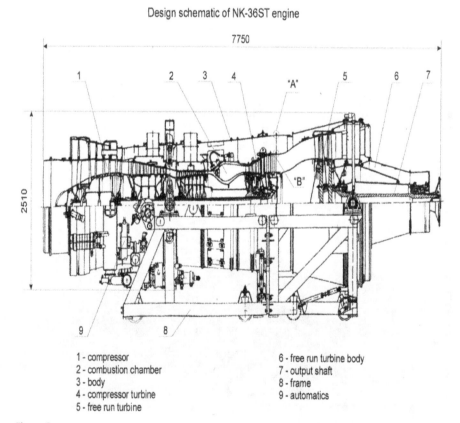

1 - compressor
2 - combustion chamber
3 - body
4 - compressor turbine
5 - free run turbine

6 - free run turbine body
7 - output shaft
8 - frame
9 - automatics

**Figure 9.**

Gas dynamic estimation of the capability of the HK-36CT engine of GDL diffuser start support is shown below.

As it was already indicated, the basic adaptation of the HK-36CT gas-turbine engine is the elimination of design structure of the free running turbine, that allows to improve characteristics from the point of view of possible application in the laser installation, for which it is necessary to bleed air with the consumption $m_B$ = 16,25 kg\sec by pressure P $\approx$22 kg / cm$^2$ and temperature T = 720K.

Reasonable temperature for a GDL gas flow channel is T = 1600K at combustion-chamber pressure of 20 kg / cm$^2$.

The given temperature is realized in the combustion chamber for fuel kerosene - air at mass ratio $K_K$ = 38,5, that is at kerosene flow rate $m_n$ = 0,422 kg / s.

In GDL working section it is necessary to ensure static temperature of combustion products Ta = 370 K, that is realized velocity exhaust of combustion products through supersonic nozzles with velocity Wà = 1700 m\s at a Mach number Ma = 4,46, isoentropy index $\kappa$ = 1,333, pressure Pa = 0,057 kg / cm$^2$.

For realization in GDL flat working section shock wave free flow of the of combustion products in atmosphere is necessary to utilize the flat aft supersonic diffuser. One of major parameters of functioning of the diffuser in GDL operation conditions is the start pressure.

The start pressure (Pst) is determined on intensity of a direct shock wave relation:

$$\frac{Pst.}{P_H} = \frac{g(\frac{1}{\lambda a})}{g(\lambda a)};$$

Where:      Pn            -   pressure behind the diffuser;

$\lambda a = \dfrac{Wa}{a_*}$      -   velocity factor equal to the ratio of exhaust velocity in

working section of GDL (Wa) to the critical velocity of sound (a ·).

The value of a velocity factor $\lambda a$ is connected to a Mach number by a ratio

$$\lambda a = \sqrt{\frac{X+1}{2} \cdot \frac{Ma}{\sqrt{1 + \frac{X-1}{2}Ma^2}}}$$

and then to a Mach number Ma = 4,46 corresponds $\lambda a = 2,32$.

The function $g = \dfrac{Fkr.}{Fa}$ determines the ratio of the geometrical area of critical cross-section of the diffuser to the area of gases exhaust and is according to expression

$$g(\lambda a) = (\frac{X+1}{2})\frac{1}{X-1} \cdot \lambda a(1-\frac{X-1}{X+1}\lambda a)\frac{1}{X-1}$$

For direct shock wave the following ratio is valid

$$\lambda a \cdot \lambda_{n.c.} = 1;$$

Where:        $\lambda_{n.c.}$   -   dimensionless velocity factor after front of a direct shock wave

Thus, the ratio of pressure of start to pressure behind the diffuser is equal

$$\frac{Pstor.}{Pn} = 13,7.$$

The wind tunnel pressure of start is usually above their steady state operating pressure up to 10-30 %.

At the same time, in GDL laser conditions, when there is a gas volume involved in a working section (region of resonator mirrors) pressure of start of the diffuser can is still increasing to the value, at which the diffuser could not started.

Hence, it is necessary to lower exhaust pressure from the diffuser, that is possible to realized through an ejection of a gas jet, flowing out from the diffuser, by a gas jet flowing out after the turbine of the engine (Fig.10).

GDL exit gas flow ejection

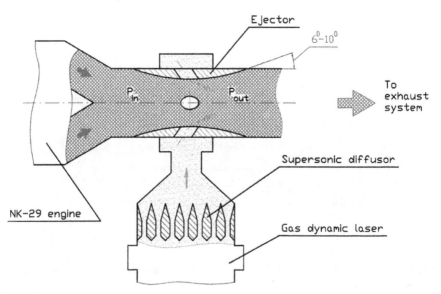

Figure 10.

The designs of similar ejectors with maximum pressure recovery are known at high speeds of flow of gas in minimum cross-section. Such ejectors designs are characterized by smooth profile of an inlet part and with small cone angles (L) on an exit, which make value L = 6 – 10 degree.

The pressure recovery coefficient for such ejectors can reach value $\dfrac{Pexit}{Pinlet} = 0,955 \div 0,965$

and even higher.

Thus the velocity factor value in minimum cross-section should be

$$\lambda = 0,75 - 0,85.$$

Thus, such ejector inlet pressure should be Pinlet. = 1,076 kg / cm², which can be provided with the gas-turbine engine without free running turbine. (Pexit. = 1,033 kg / cm²).

At a velocity factor $\lambda = 0,8$ in minimum cross-section and ejector inlet pressure Pinlet. = 1,076kg / cm² the static pressure in minimum cross-section becomes equal Pmin =0,73kg / cm², that allows to organize introduction of flowing out from the GDL diffuser gas in area of minimum cross-section of the exhaust device of the gas-turbine engine and by that to lower pressure behind the diffuser GDL at least down to Pexit = 0,8 instead of 1,033.

Such reduction of pressure will allow to lower start pressure of the GDL down, to a reasonable level of 18-20 atm.

### 4.4. Design solution for GDL power unit on the basis of the gas-turbine engine

In view of the technical plan generated by the authors of the present technical proposal, experts of the Samara complex «Motorostroitel», has developed on the basis of the HK-36CT gas-turbine engine the engine for a laser complex under index HK-29.

The HK-29 engine (Fig.11) represents the gas-turbine engine obtained as a result of adaptation of the engine HK-36CT, which consists of the following:

### 4.4.1. Dismantling of the free running turbine with the output shaft

The dismantling is made through the joint "A" flange of a engine body (see Fig. 4), through the joint "B" of the hot gas pipe down to the turbine of the compressor and finally the free running turbine together with a turbine casing is removed from the engine. As the frame, on which the engine is mounted, consists of two not connected permanently sections, the removal of the second turbine side section does not represent any complexity.

For organization of the smooth gas flow from the modified engine, and also for a fixing of the exhaust system to the engine, two conical adapters (casing) and internal fairing made of a stainless steel are added.

NK-29 engine

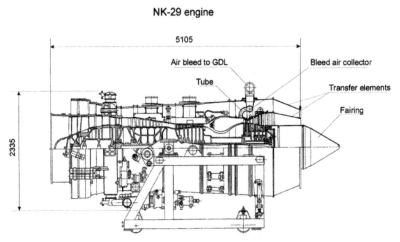

**Figure 11.**

The external adapter is terminated by a flange to be fixed with the engine ejector. The adapters and fairing were designed by «Motorostroitel», supplier in accordance to the aerodynamic characteristics of a gas stream and the strength requirements.

Ejector, which scheme is shown in Fig.9, is not the engine item. The ejector design is calculated and finally developed on the basis of engine, and laser complex exhaust system parameters and GDL exit characteristics (down the supersonic diffuser).Ejector design represents a pipe from high-temperature steel having symmetrical narrowing in the middle part ahead of critical cross-section. The pipe can be manufactured from three parts with flange or welded connections.

The middle narrowing part can be turned from a thick-walled billet or from a rolled ring and has mating collars for different side collector for convenience of assembly for welding. The orifices connecting cavities of exhausts of the engine and GDL are located around the perimeter of critical cross-section. The collector is manufactured by stamping of a sheet, has the welded branch pipe and serves for uniform distribution of gas after GDL on all cross-section of an ejector. The adaptation of the engine with removal of the free running turbine considerably reduces overall dimensions and weight of the installation. For clarity of estimation of advantage in overall dimension reduction of the modified engine the presentations of Figs. 8 and 10 are made in the same scale.

### 4.4.2. Air bleeding to GDL

The next considered problem is the air bleeding after the compressor to GDL.

During our long searching of the engineering solution for the engine adaptation the aviation specialists have offered to realize the air bleeding from the exit of the combustion chamber second contour, not from the compressor final stage exit, i.e. from a channel of a cooling flow of the combustion chamber. This solution is connected to the fact, that combustion chamber of the gas-turbine engine is one of the most thermal stressed items of the engine, and the cooling parameters are results of long development stage. Therefore the air bleed behind the compressor, reducing the air flow of the combustion chamber cooling flow, will change steady state thermal status, which can result in undesirable consequences for engine operation.

## GDL air bleed block design

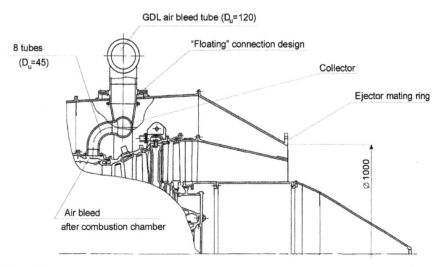

**Figure 12.**

The collector ring for air bleeding with a branch pipe to GDL supply was recommended to position in a zone of the exit of cooling flow of the combustion chamber, and for that the chamber and collector are connected by angular branch pipes (Fig.11).The angular branch pipes, 8 pieces in quantity, can be manufactured of a pipe or stamped sheet from two halves and have leak tight flanges to cooperate with the envelope of chamber cooling flow channel. The angular branch pipes are fixed by welding to the collector ring, which has the terminal flanged tap. To this tap the branch pipe to supply air to GDL is fixed. This branch pipe passes through an outer shell of a engine body and has "floating" attachment to it excluding deformation of designs at a thermal expansion of contacting elements.

It exists from the experience of such bleed scheme, however particular design should pass a number of development tests with the purpose of experimental selection of channel pressure reduction to ensure calculated bleed air flow rate. All other elements of the engine remain without change, including all start and cutoff automatics and operation control.

Thus engine HK-29, obtained as a result of adaptation of the engine HK-36CT will require minimum development, and manufacturing costs.

The basic performances of the engine HK-29. All parameters of the engine are indicated for a nominal mode.

| | |
|---|---|
| Fuel | - Kerosene TC-1 |
| Power, MW | - 25 |
| Air flow, kg / s | - 107,2 |
| Compressor pressure ratio, Pk | - 23,12 |
| CC gas temperature, K | - 1420 |
| Gas exit temperature, K | - 698 |
| Propellant consumption, kg / s | - 1,434 |
| Engine weight, with frames, kg | ~- 5000 |

The engine HK-29 is single mode, as by activity of the laser installation is required neither decrease, nor increase of the power, that provides reliability and life value growth.

The engine HK-29 should be manufactured by aviation plant irrespective of, whether it is refurbished through adaptation of the engine HK-36CT or again assembled with use of furnished engine parts from production engines.

Thus, as a result of the carried out research work not only the optimistic results for possibility of adaptation of the gas-turbine engine for GDL gas dynamic complex were obtained, but also design activities by definition of general realization of such engine are executed in practice. And, what is very valuable, this activity is executed by the experts of aero-engines developing - producing company.

## 5. The mobile laser installation

In view of all computational investigations, design studies and engineering estimations within the present paper the provisional structure of the autonomous laser installation for clearing of surfaces of reservoirs of petroleum recordings is produced. The general view of laser installation is shown in Fig.13.As it can be seen from the figure, the installation represents modular assembly. The engine has an own fixing frame, autonomous subsystems and producer - supplier. GDL also represents the complete aggregate, the developer and manufacturer of which is not connected to other suppliers. The same concerns are valid for the fuel storage system (is not shown in figure).Task of the installation developer as a whole is the coordination of these units both on arrangement on a vehicle, and organization of interaction during operation. For example, relative arrangement of the engine and laser can be arbitrarily, depending on overall dimensions and device of a vehicle, however in any case the fulfillment of following conditions is necessary:

• The engine and GDL should not have of rigid connection on the frame, the GDL frame should have vibration absorption plates.

- The engine and GDL should be placed as close as possible, as the unreasonable elongation of gas lines connecting them, will result in to additional losses because of pressure reduction.
- On fire protection control requirement fuel storage container should be placed on distance from the engine and GDL, and is separated from them by a fire-prevention wall.
- For protection of the attendants the remote control panel should be born for limits of the installation, and input part of the engine and line of a beam are protected by special fences. The control of a scanning mirror also is carried out for limits of the installation.
- The direction of exhaust gases should not intersect structural items of the installation.

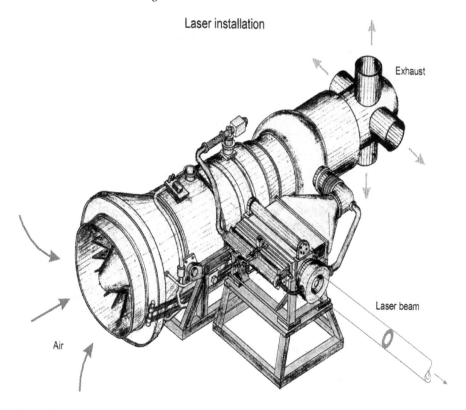

**Figure 13.** Laser installation.

As the development of the laser installation does not assume development of testing site for development testing, prior to deliveries the development installation should be assembled, on which all designer and operational problem solutions will be found.

## 6. Conclusion

The Purpose of presented in the paper project is to develop a laser technology of cleaning of large areas of seas and other water surfaces from oil film contamination. This paper also aims development of composition scheme, technological requirements and engineering design solution of aggregates and units of vehicle, vessel or aircraft compatible mobile GDL (100 - 250 kW) intended to solve this important problem of environment protection. This method proposed for development is expected to complement other traditional methods, which usually more successfully treat bulk layer oil pollution but do not match to eliminate up to 100 μ oil films, the latter usually being spread over numerous square kilometers of sea and kilometers of seacoasts. Note about one million tons of such films drift now in the World Ocean, as consequence of such disasters.

Thus, this promising approach has a fundamental science basis, from one hand, and skill and experience of missile jet specialists and their laser prototypes, from another hand. The investigations are planned to be accomplished in three basic directions:

1.  theoretical and experimental research of the laser radiation action on thin oil and it's derivatives films over water surface, evaporation in three basic directions:
2.  experimental optimization of laser optical scheme and operation modes in order to achieve the maximum optical and gas - dynamic efficiency.
3.  engineering design, composition scheme and accompanied problems solution in order to adapt the laser module for vehicle, ship or aircraft transportation.

Results expected are as following:

-   Certain recommendations on oil films elimination from large water areas, including conditions of real tanker and pipeline disaster, by means of laser radiation; development of technologies of oil and it's derivatives films burning out (mode number one) and oil films sucking up and saving of oil products for future efficient usage (mode number two) by means of high power mobile $CO_2$- GDL.
-   Development of self-contained mobile $CO_2$ GDL tailored for these environmental problems solution.

These investigations may result in significant progress in the following branches of science:

a.  physics of liquid inhomogeneous films, phenomena on the boundary of water-carbohydrate composition, phase transitions and chemical processes in two-phase or multi-phase compositions under high power $CO_2$ laser radiation and related phenomena;
b.  problem of active resonators for high power GDL, different methods of laser efficiency calculation and simulation, control of temporal radiation modulation and optimization of performance, with respect to the problem of beam quality.

The author has completed the conceptual analysis of GDL installation development for disposal of petroleum films from a water surface. The basic capability of development of such installation is shown. The basic characteristic calculations of the system are made, the schematic and design solutions of basic installation are presented.

Besides of that, the GDL bench has been restituted by the author of the paper and his colleagues on their own costs and some demonstration operations with scanning the beam over water surface covered with petroleum film were carried out. The experimental confirmation of film effective burning mode and mode of oil film gathering was obtained. The next part of realization phase can be the development of the installation preliminary design, in case of the customer objective and specific technical tasks, formulated by him. The preliminary design will include complete study of a design of all elements of the installation, technological works, will define all suppliers of the developing systems and sub suppliers from obtaining from them of the consent on terms and conditions. In the preliminary design, schedules of all phases of installation development will be worked out: issue of the working documentation, technology modification, manufacturing of a prototype, prototype development and beginning of installations deliveries.

Thus, according to the estimation the experimental mobile CO2 GDL installation for water surface cleaning can be designed, manufactured and tested for not longer than two - two and half years.

## Author details

V. V. Apollonov

*Prokhorov General Physics Institute of RAS, Moscow, Russia*

## Acknowledgement

The author would like to acknowledge the valuable contributions made to realization of this work **"JET ENGINE BASED MOBILE GAS DYNAMIC CO2 LASER FOR water SURFACE CLEANING"** by V.V. Kijko, Yu.S. Vagin, and A.G. Suzdal'tsev.

## 7. References

[1] V.V. Apollonov, A.M. Prokhorov, "Universal laser for industrial, scientific and ecological use", Proceedings of GCL/HPL Conference, St. Petersburg, p.140, 1998.

[2] V.V. Apollonov, "Ecologically safe High power Lasers", Proceedings of Lasers-2001 Conference, p.3, 2001;

[3] V.V. Apollonov, "High power autonomous CO2 GDL (100kW,CW/P-P modes) for new technologies development and environment protection"; Russia –NATO Int. Seminar, Moscow, 2005, Proceedings of GCL/HPL Conference, Lisbon, p.35,2008.

[4] V.V. Apollonov "New application for high power high repetition rate pulse-periodic lasers", "Laser pulse Phenomena and Applications" Intech,p.19,2010.

[5]  V.V. Apollonov, "Oil films elimination by laser", Oboronzakaz, №17, December, p. 33, 2007.
[6]  V.V. Apollonov, Yu, S. Vagin, V.V. Kijko, "High rep.rate P-P lasers", Patent RF № 2175159

# Seeing Invisible

# Chapter 5

# Ultrashort Laser Pulses for Frequency Upconversion

Kun Huang, E Wu, Xiaorong Gu, Haifeng Pan and Heping Zeng

Additional information is available at the end of the chapter

## 1. Introduction

Single-photon frequency upconversion is a nonlinear process where the frequency of the signal photon is translated to the higher frequency with the complete preservation of all the quantum characteristics of the "flying" qubits [1,2]. One promising application of the single-photon frequency upconversion is converting the infrared photons to the desired spectral regime (usually visible regime) where the high performance detectors are available for sensitive detections [3-6]. The infrared single-photon frequency upconversion detection technique has been successfully used in a variety of applications, including infrared imaging [7] and infrared ultra-sensitive spectroscopy [8,9]. Additionally, such upconversion detector has greatly benefited the applications stringently requiring efficient photon detection in optical quantum computation and communication [10-12]. Furthermore, due to the phase-matching requirement in the frequency upconversion process, the upconversion detectors have some unique features, such as narrow-band wavelength acceptance [13] and polarization sensitivity [14], both of which render them very useful for fiber based quantum systems [15]. In the recent decade, experiments bear that the photon correlation, entanglement, and photon statistics are all well preserved in the coherent frequency upconversion [16-18].

Although the demonstration of frequency upconversion technology for strong light could be dated to the late 70s of last century [19], the single-photon frequency upconversion was just achieved at the beginning of this century [20]. The advent and rapid evolution of fabricating the periodically reversed nonlinear media led to the widespread use of quasi-phase matching (QPM), which has opened up new operating regimes for nonlinear interactions [21]. Rapidly growing interest has been focused in recent years on proposing novel schemes for achieving single-photon frequency upconversion with high efficiency and low noise and expanding its applications in all-optical nonlinear signal processing and in quantum state manipulation. With the help of extracavity enhancement, highly efficient up-conversion at single-photon level has been demonstrated by using bulk periodically poled lithium niobate (PPLN) crystals

[3]. By means of intracavity enhancement, a stable and efficient single-photon counting at 1.55 μm was achieved without the requirement of sensitive cavity servo feed-back [4,22]. Recently, frequency upconversion based on the PPLN waveguide attracts more and more attention [9,23]. All of the above single-photon frequency upconversion systems were pumped by the strong continuous wave (CW) lasers which would inevitably bring about severe background noise due to parasitic nonlinear interactions. Frequency upconversion based on pulsed pump arises to provide an effective solution [24]. To make sure that every signal photon can be interact with the pulsed pump field, synchronous pumping frequency upconversion system is thus introduced, which greatly improves the total detection efficiency. VanDevender *et al.* reported a frequency upconversion system based on electronic synchronization of the pump source and signal source with a repetition rate of just 40 kHz [5]. For satisfying applications of high-speed single-photon detection, fast and efficient single-photon frequency upconversion detection system operating at tens of MHz was realized based on the all-optical synchronized fiber lasers [25-27].

Thanks to the ultrashort optical pulses, intense peak power to obtain unit conversion efficiency could be achieved with a modest average pump power, thus loosening the restriction for the available output power from the pump laser [3,4]. Additionally, the background noise induced by the strong pump can be effectively reduced due to the pulsed excitation. Thus ultrashort pulses constitute ideal means as the pump source of frequency upconversion. The pulsed pumping system could not only permit efficient photon counting of infrared photons, but also quantum manipulation of single photons. One can realize a quantum state router by having a control on the multimode pump source [28]. Current researches also demonstrate simultaneous wavelength translation and amplitude modulation of single photons from a quantum dot by pulsed frequency upconversion [29]. Additionally, the concept of a quantum pulse gate is presented and an implementation is proposed based on spectrally engineered frequency upconversion [30].

In this chapter, we review the recent experimental progress in single-photon frequency upconversion with synchronous pulse pumping laser. The signal photons were tightly located within the synchronous pump pulses [25]. For improving the conversion efficiency, the specific control of the synchronized pulses was required, which led to the development of temporally and spectrally controlled single-photon frequency upconversion for pulsed radiation [26,27]. The compact fiber-laser synchronization system for fast and efficient single-photon frequency upconversion detection is of critical potential to stimulate promising applications, such as infrared photon-number-resolving detector (PNRD) [18,31] and ultrasensitive infrared imaging at few-photon level. This chapter is organized as follows. After this brief introduction, we present the basic theory for frequency upconversion in quantum frame in Section 2. Section 3 presents the experimental realization of synchronous pulsed pumping. Applications based on the synchronous pumping frequency upconversion system will be highlighted in Section 4 by the examples of infrared photon-number-resolving detection and several-photon-level infrared imaging. We conclude the chapter in Section 5 by emphasizing that the synchronous pumping frequency upconversion system will benefit numerous applications not only in infrared photon counting and also in manipulation of the quantum state of single photons.

## 2. Quantum description of frequency upconversion

Frequency upconversion based on sum frequency generation (SFG) is an optical process by which two optical fields combine in a quadratic nonlinear medium to generate a third field at a frequency equal to the sum of the two inputs as shown in Fig. 1. The theory of the SFG is well established for a long time and applied in a variety of classical and quantum optical applications. In this section, we go through the fundamental theory of SFG and derive the important results in quantum architecture. We investigate the quantum features according to different pumping methods for frequency upconversion, respectively. Specially, we will concentrate on the detailed discussion of the quantum characteristics in the multimode pumping scheme.

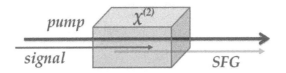

**Figure 1.** Schematic illustration of the frequency upconversion process.

### 2.1. Single-mode frequency upconversion

In quantum optics, the frequency upconversion is a nonlinear optical process, in which a photon at one frequency is annihilated and another photon at a higher frequency is created. We start with a Hamiltonian for three-wave mixing:

$$\hat{H}_{3W} = i\hbar g(\hat{a}_p \hat{a}_1 \hat{a}_2^\dagger - H.c.), \tag{1}$$

where $\hat{a}_1$ and $\hat{a}_p$ is the annihilation operators corresponding to the signal photon at $\omega_1$ and pump photon $\omega_p$, respectively. $\hat{a}_2^\dagger$ is the creation operator corresponding to the upconverted photon $\omega_2$, $g$ is the coupling constant which is determined by the second-order susceptibility of the nonlinear medium, and $H.c.$ denotes a Hermitian conjugate. If the pump field is very strong with negligible depletion, as in the case here, we can classically treat it as a constant $Ep$. So the Hamiltonian in Eq. (1) becomes

$$\hat{H} = i\hbar g E_p(\hat{a}_1 \hat{a}_2^\dagger - H.c.). \tag{2}$$

Since for most cases, the upconverted frequency wave was in vacuum state at the input of the nonlinear medium. So the initial condition at the input facet of the nonlinear medium could be written as

$$|\Phi\rangle = |\Psi_1, 0_2\rangle, \tag{3}$$

where $|\Psi_1\rangle$ represents the input signal state and $|0_2\rangle$ represents the vacuum input of the SFG state. The dynamics of the input and output quantum fields in phase-matched SFG process can be described by the coupled-mode equations as

$$\hat{a}_1(L) = \hat{a}_1(0)\cos(|gE_p| L) - \hat{a}_2(0)\sin(|gE_p| L), \tag{4}$$

$$\hat{a}_2(L) = \hat{a}_2(0)\cos(|gE_p| L) + \hat{a}_1(0)\sin(|gE_p| L), \tag{5}$$

where $L$ is the interaction length in the nonlinear crystal. The corresponding creation operators can be found by taking the Hermitian conjugates of these equations.

Eq. (5) indicates that frequency translation of any quantum state at $\omega_1$ to the same quantum state at $\omega_2$ with unity efficiency is possible, even at the single-photon level, if $|gE_p| L = \pi/2$. Note that the coherent quantum transduction from one frequency to a higher frequency would preserve complete coherence properties including entanglement, quantum correlation, and photon statistics.

At the output of the nonlinear medium, the average photon number of infrared signal can be calculated from expected value of the photon number operator:

$$\begin{aligned} N_1(L) &= \left\langle \hat{a}_1^\dagger(L)\hat{a}_1(L) \right\rangle \\ &= \left\langle \psi_1 \left| \hat{n}_0 \right| \psi_1 \right\rangle \cos^2(|gE_p| L) \\ &= N_1^0 \cos^2(|gE_p| L), \end{aligned} \tag{6}$$

where $\hat{n}_0$ denotes the photon number operator of the input infrared signal; $N_1^0$ is the average photon number of the input signal.

Meanwhile, the average number of the SFG photons can be likewise given by

$$N_2(L) = N_1^0 \sin^2(|gE_p| L). \tag{7}$$

From Eq. (6) and Eq. (7), the upconversion efficiency can be given as

$$\eta = \frac{N_2(L)}{N_1(0)} = \sin^2(|gE_p| L). \tag{8}$$

Additionally, we can get $N_2(L) + N_1(L) = N_1^0$, indicating the energy conservation during the conversion process. The correlation between the upconverted photons and unconverted photons is shown in Fig. 2.

The joint probability $P_{12}$ of simultaneously detecting a photon at both $\omega_1$ and $\omega_2$ at the output of the frequency upconverter is proportional to $\langle \hat{n}_1 \hat{n}_2 \rangle$. With the help of Eq. (4,5) we readily obtain that

$$\begin{aligned} \langle \hat{n}_1 \hat{n}_2 \rangle &= \left\langle \hat{a}_1^\dagger(L)\hat{a}_2^\dagger(L)\hat{a}_2(L)\hat{a}_1(L) \right\rangle \\ &= [\langle \hat{n}_0^2 \rangle - \langle \hat{n}_0 \rangle]\cos^2(|gE_p| L)\sin^2(|gE_p| L). \end{aligned} \tag{9}$$

Inserting Eq. (8) into the above equation, we have

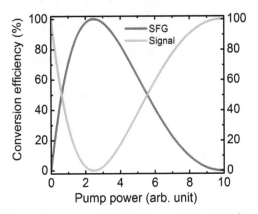

**Figure 2.** Correlation between the upconverted SFG photons and unconverted signal photons.

$$\langle \hat{n}_1 \hat{n}_2 \rangle = [\langle \hat{n}_0^2 \rangle - \langle \hat{n}_0 \rangle ] \eta (1 - \eta), \tag{10}$$

which is dependent on the conversion efficiency of the frequency upconversion process.

Moreover, the intensity cross-correlation function $g^{(2)}(\tau)$ at $\tau = 0$ is then obtained from

$$g^{(2)}(0) = \frac{\langle \hat{n}_1 \hat{n}_2 \rangle}{\langle \hat{n}_1 \rangle \langle \hat{n}_2 \rangle} = \frac{\langle \hat{n}_0^2 \rangle - \langle \hat{n}_0 \rangle}{\langle \hat{n}_0 \rangle^2}. \tag{11}$$

When input signal photons are in single photon state, meaning $\langle \hat{n}_0^2 \rangle = \langle \hat{n}_0 \rangle$, the probability of detecting a photon both in frequency $\omega_1$ and $\omega_2$ is then zero. The unconverted infrared photons and SFG photons are anti-correlated. When the incident photons are in a coherent state, meaning $\langle \hat{n}_0^2 \rangle - \langle \hat{n}_0 \rangle = \langle \hat{n}_0 \rangle^2$, the intensity cross-correlation equals to 1, which means no anti-correlation will be observed. It indicates that the frequency upconversion process is a random event for the individual incident signal photons.

All above results are obtained on the basis of perfect phase matching. The phase matching condition is sensitive in the SFG interaction due to the limit spectral bandwidth of the nonlinear medium. Taking into account the phase mismatching [32], the quantum state of the SFG photons can be expressed as

$$\hat{a}_1(L, \Delta\omega) = \cos(qL)\exp(-i\Delta\omega\Delta kL/2)\hat{a}_1(0, \Delta\omega)$$

$$- \frac{igE_p}{q}\sin(qL)\exp(-i\Delta\omega\Delta kL/2) \tag{12}$$

$$\times [\frac{\Delta\omega\Delta k}{2gE_p}\hat{a}_1(0, \Delta\omega) + \hat{a}_2(0, \Delta\omega)],$$

$$\hat{a}_2(L,\Delta\omega) = \cos(qL)\exp(-i\Delta\omega\Delta kL/2)\hat{a}_2(0,\Delta\omega)$$

$$+\frac{igE_p}{q}\sin(qL)\exp(-i\Delta\omega\Delta kL/2) \qquad (13)$$

$$\times[\frac{\Delta\omega\Delta k}{2gE_p}\hat{a}_2(0,\Delta\omega) + \hat{a}_1(0,\Delta\omega)],$$

where $\Delta k$ represents the phase-mismatch at frequency detuning $\Delta\omega$, and $q = [(gE_p)^2 + (\Delta\omega\Delta k/2)^2]^{1/2}$. At zero detuning $(\Delta\omega = 0)$, Eqs. (12) and (13) can be deduced to Eqs. (4) and (5), respectively. The unity single-photon frequency conversion can be achieved when $| gE_p | L = \pi/2$ is fulfilled. At nonzero detuning $(\Delta\omega \neq 0)$, although a similar condition $qL = \pi/2$ can be fulfilled, the perfect quantum conversion is impossible due to the disturbance of the vacuum state $\hat{a}_2(0,\Delta\omega)$. Therefore, for practical experiments, the spectral requirement of the pump field for frequency upconversion should be well concerned.

## 2.2. Synchronous pumping frequency upconversion

Could the complete quantum state transduction be feasible when the signal and the pump source were both in the multi-longitudinal modes? This is the instance of the synchronous pumping frequency upconversion system. Obviously, the frequency conversion process becomes more complicated due to the commutative nonlinear coupling of the pump modes and signal modes.

The Hamiltonian can be rewritten as

$$\hat{H} = i\hbar g \sum_{ij} E_{pi}(\hat{a}_{1j}\hat{a}_{2ij}^\dagger - H.c.), \qquad (14)$$

where the $E_{pi}$ is the pump electric field related to each longitudinal mode numbered by $i$, and the $\hat{a}_{1j}$ is the annihilation operator of the infrared signal photons related to each longitudinal mode numbered by $j$, the $\hat{a}_{2ij}^\dagger$ is the creation operator of the SFG photons corresponding to the longitudinal mode of both pump field and infrared signal photons. Then the dynamics of the input and output quantum fields in phase-matched SFG processes can be described by the coupled-mode equations as

$$\frac{d\hat{a}_{1j}}{dt} = \frac{1}{i\hbar}[\hat{a}_{1j},\hat{H}] = -g\sum_i E_{pi}\hat{a}_{2ij}$$

$$\frac{d\hat{a}_{2ij}}{dt} = \frac{1}{i\hbar}[\hat{a}_{2ij},\hat{H}] = gE_{pi}\hat{a}_{1j}. \qquad (15)$$

The superposition state of the infrared signal photons and SFG photons are represented by the operators $\hat{a}_1$, $\hat{a}_2$, which are defined as

$$\hat{a}_1 = \sum_i \hat{a}_{1j}$$

$$\hat{a}_2 = \sum_j \sum_i C_i \hat{a}_{2ij}, \qquad (16)$$

where $E_p^2 = \sum_i E_{pi}^2$ and $C_i = E_{pi} / E_p$ denotes the probability amplitude of each longitudinal mode. Then the coupled-mode equation can be trivially solved by using the initial condition at the input facet of the nonlinear medium to yield

$$\hat{a}_1(L) = \hat{a}_1(0)\cos(|gE_p|L) - \hat{a}_2(0)\sin(|gE_p|L), \tag{17}$$

$$\hat{a}_2(L) = \hat{a}_2(0)\cos(|gE_p|L) + \hat{a}_1(0)\sin(|gE_p|L). \tag{18}$$

We get the same results as that in a single-longitudinal mode situation (Eqs. (4, 5)). When $|gE_p|L = \pi/2$ is satisfied, the coherence properties of the incident signal photons could be maintained during the upconversion process in this multi-mode system.

Here we can clearly see that either in single-mode regime or in multi-mode regime, the upconversion can, in principle, be used to transduce one photon at a given wavelength to another wavelength in preservation of all the quantum characteristics. Such quantum upconverter would not only be useful in the fields related to efficient infrared photon counting, but also helpful to implement novel quantum functions such as quantum interface and quantum gate or quantum shaper. In next section, experimental realization of the single-photon frequency upconversion with synchronously pumping will be demonstrated at length.

## 3. Experimental realization of single-photon frequency upconversion system

To achieve efficient and low-noise single-photon frequency upconversion system, various schemes have been proposed by the researchers for different applications in recent years. The upconversion technique typically requires a sufficiently strong pump to achieve unity nonlinear frequency conversion efficiency in a quadratic nonlinear crystal. The requisite strong pump can be achieved by using an external cavity or intracavity enhancement [3,4] or a waveguide confinement [9,23]. With such high intensity of pump, frequency up-conversion was implemented with almost 100% internal conversion efficiency. Nevertheless, a strong pump field inevitably brings about severe background noise because of parasitic nonlinear interactions, such as spontaneous parametric downconversion and Raman scattering. Pulsed pumping technique is given rise to circumscribe induced noise within the narrow temporal window of the pump pulses. Ultralow background counts of 150/s was reported with the help of a long-wavelength pump scheme [24]. In order to include every signal photon within the pump pulse for improving the total detection efficiency, researchers recently develop a coincidence single-photon frequency upconversion system [5,25-27]. In this section, we will focus on the synchronous pulsed pumping technique and its experimental results.

Fig. 3 shows the experimental setup of the synchronous single-photon frequency upconversion detection. The system was composed of a passive master-slave synchronization fiber-laser system and a single-photon frequency upconversion counting system. In the synchronization fiber-laser system, both two fiber lasers were passively mode-locked by the nonlinear polarization rotation in the fiber cavity, operating at the repetition rate of 17.6 MHz to

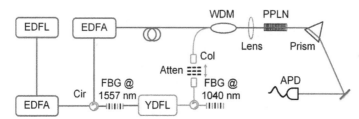

**Figure 3.** Experimental setup of the synchronous single-photon frequency upconversion detection. YDFL, ytterbium-doped fiber laser; EDFL, erbium-doped fiber laser; EDFA, erbium-doped fiber amplifier; Cir, circulator; Col, collimator; Atten, attenuator; WDM, wavelength-division multiplexer; FBG, fiber Bragg grating; PPLN, periodically poled lithium niobate crystal; APD, avalanche photodiodes.

satisfy the high-speed detection. The master laser was an Er-doped fiber ring laser. The ring cavity consisted with 1.5-m Er-doped fiber, standard single-mode fiber and dispersion-shifted fiber. By optimizing the lengths of the single-mode and dispersion-shifted fibers while maintaining the total cavity length, the dispersion in the cavity was well controlled. The spectrum of the output laser centered at 1563.8 nm with a full width at half maximum (FWHM) of 6.2 nm. A narrow spectral portion of the master laser was extracted by the FBG reflection for approaching the QPM bandwidth of the PPLN. This part was further amplified by an Er-doped fiber amplifier (EDFA) as the pump source. The maximum output power was about 60 mW and the output spectrum from the amplifier centered at 1557.6 nm with an FWHM of 0.56 nm [Fig. 4(a)]. The transmission from the FBG was injected as a seed to the slave laser cavity via a 1040/1557-nm wavelength-division multiplexer (WDM) to trigger the mode-locking of the Yb-doped fiber laser. An FBG was used as well at the slave laser output to control the spectral matching of the signal photons. The filtered spectrum was centered at 1040.0 nm with the bandwidth about 0.35 nm (FWHM) [Fig. 4(b)]. Injection-locking of the slave laser ensured the synchronization between the master and slave laser due to the fact that the transmission from the FBG shared exactly the same spectral mode separation with the reflection.

With this master-slave configuration, the two synchronized lasers could be isolated from each other, and thus free from mutual perturbation. This all-optical synchronization technique is superior to other passive synchronization methods due to its simplicity and robustness [33-35]. If self-started mode-locking existed in the slave laser, synchronized injection-locking could be obtained as well, but only within a sensitive slave cavity length match and could not last long, which would not be implemented in the frequency upconversion system. The self-started mode-locking would compete with the injection-triggered mode-locking, leading to the instability. To reduce the instability, self-started mode-locking was avoided in the slave fiber laser by properly adjusting the laser polarization state in the cavity. In such a critical position, the slave laser could only be mode-locked with the presence of the seed injection. The cavity length of the slave laser was adjusted to be the same as the master laser by carefully moving one of the fiber collimators on the stage. Eventually, the slave laser was synchronized with the master laser at the same repetition rate. And the stability of synchronization was enhanced by optimizing the master laser injection polarization. With such setup, we achieved the maximum cavity mismatch tolerance of 25 μm with the long-term stability of several hours.

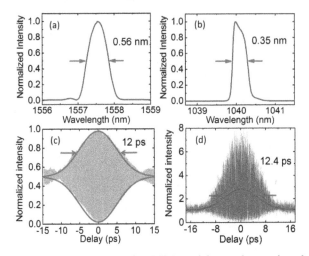

**Figure 4.** Spectrum of the pump source from the EDFL (a) and the signal source from the YDFL (b), respectively. Autocorrelation pulse profiles of the pump source (c) and the signal source from (d), respectively. Solid symbols are the experimental data and solid curves are the Gaussian fits to the data.

The signal pulse duration was measured by an autocorrelator with the FWHM bandwidth of 12 ps [Fig. 4(c)], corresponding to the actual pulse duration of 6 ps by taking the denominator of 2. In order to optimize the upconversion efficiency, the pump pulse duration relative to the signal pulse needs to be considered carefully [5]. The total conversion efficiency dependent on the pulse overlapping is given by

$$P_{overlap} = \int_{-\infty}^{+\infty} P_0(I_p(t))I_s(t)dt, \tag{19}$$

where $P_0$ is the probability of upconversion dependent on pump intensity $I_P$ as $\sin^2(I_p^{1/2})$, and $I_s(t)$ is the normalized input pulse profile

$$\int_{-\infty}^{+\infty} I_s(t)dt = 1. \tag{20}$$

The simulated conversion efficiency shown in Fig. 5 indicates that the total efficiency increases as the pump pulse duration goes longer by assuming a constant FWHM duration (6 ps) for the signal pulse. When the pump pulse duration is much shorter than the signal pulse, the conversion efficiency is quite low as most of the photons distribute outside of the pump pulse temporal window. On the other hand, too long pulse duration will lead to reduction of the energy utilization efficiency and increase of the background counts. Therefore, to achieve a conversion efficiency over 90% with experimentally available pump intensity, it is theoretically necessary to make pulse duration ratio between the pump and signal slightly larger than 1.2.

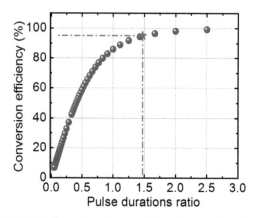

**Figure 5.** Simulated conversion efficiency as a function of the pulse duration ratio between the pump and signal. The red star indicates the experimental situation.

Thanks to the spectral filtering by the FBGs in the experiment, the pulse durations of the master and slave lasers were stretched according to the time-frequency Fourier transform. With the help of cavity dispersion management in the master laser cavity, the desired temporal match was achieved. The pump pulse duration was measured to be 8.8 ps [Fig. 4 (d)]. In this way, the signal photons and pump pulses were well matched temporally, guaranteeing an efficient upconversion with a relatively low background noise. According to the theoretical simulation, the conversion efficiency can reach 95% as shown by the red star in Fig. 5.

In the experiment, the timing jitter between the signal and pump pulses was measured to be as low as 45 fs shown in Fig. 6. Such a low timing jitter would impose a negligible influence on the temporal distribution of the signal photons within the pump pulse window. So the single-photon signal could be synchronously gated with a sufficiently high stability for coincidence upconversion detection.

As shown in Fig. 3, the signal beam and pump beam were then combine by a 1040/1557-nm WDM before being focused at the center of the 50-mm-long PPLN crystal. The temperature of the crystal was optimized at 130.4 °C for the grating period of 11.0 μm. The temperature was high enough to avoid photorefractive effects [36]. The infrared object beam interacted with the pump beam, and was upconverted through SFG to generate the visible photon at 624 nm.

For a single-photon frequency upconversion system, besides efficiency and stability, background noise was also a very important metrics for the detector performance. To analyze the noise, the spectrum before the filtering system was recorded. No observable peaks around the wavelength of the sum-frequency photons appeared when no signal photons were incident, revealing that the major noise from pump-induced parametric fluorescence was almost eliminated in our experiment. Considering the pulsed pump mode, the noise was localized within a very short pump time window which was much shorter than any electronic gates applied on the APD, leading to a very low noise counts on the APD.

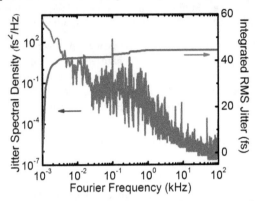

**Figure 6.** Timing jitter power spectral density and the integrated timing jitter in Fourier domain.

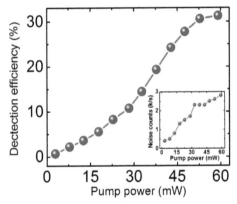

**Figure 7.** Conversion efficiency as a function of the pump power. Inset: background counts vs the pump power.

As a result, the maximum detection efficiency of 31.2% was achieved at the pump power of 59.1 mW, with the corresponding background counts of $2.8\times10^3$ s$^{-1}$ shown in Fig. 7. The maximum conversion efficiency of the system was calculated to be 91.8% after taking into account the transmittance of the filters and the quantum efficiency of the Si-APD SPCM. Since the signal and pump pulses were focusing on PPLN crystal, the imperfection of the conversion efficiency might be mainly caused by the spatial mode mismatching of the pump and signal. Compared with the background counts of CW pumping ($1\times10^5$ s$^{-1}$) [3,4], it was about two orders smaller in synchronously pulsed pumping scheme. As a figure of merit, we calculated the noise equivalent power (NEP = $h\upsilon(2R_{BC})^{1/2}/\eta$) divided by the operation rate f, where $h\upsilon$, $R_{BC}$ and $\eta$ are the energy of a signal photon, the background noise, and the detection efficiency, respectively. NEP/f was an important parameter of the sensitivity of an optical detector, especially referring to the determinant of the data acquisition time in general and the key generation rate in quantum key distribution (QKD) systems [23]. At the peak detection efficiency, the NEP/f was as low as $2.6 \times 10^{-24}$ W/Hz$^{3/2}$, thus showing that such scheme is suitable for the fast and efficient infrared single-photon detection.

The synchronous pumping single-photon frequency upconversion system is not only applied in the infrared photon detection with high efficiency and low background noise. More importantly, due to the unique quantum features in coherence preservation, it may stimulate promising applications with compact all-fiber devices in temporal and spectral control of single-photon nonlinear photonics. In Section 4, some applications are presented.

## 4. Applications in infrared PNRD and imaging

We believe that the compact fiber-laser synchronization system for fast and efficient single-photon frequency upconversion is of critical potential to stimulate other promising applications, such as high-speed QKD and quantum interface. The synchronous frequency upconversion system will also benefit infrared photon number resolving detection or few-photon-level infrared imaging.

### 4.1. Infrared photon-number-resolving detection

PNRD supports promising and important applications in few-photon detection, nonclassical photon statistics measurements, fundamental quantum optics experiments, and practical quantum information processing [37–39]. In the visible regime, PNRD with high quantum efficiency and low dark counts could be realized by employing silicon-based multipixel photon counter (Si-MPPC) [40]. Meanwhile at telecom wavelengths, PNRD could be achieved with InGaAs-based avalanche photodiodes (APD) operating in non-saturated mode [41]. However, PNRD for the wavelengths around 1 μm is still a bottleneck because both Si- and InGaAs-detectors are insensitive at those wavelengths. Recent advances in infrared photon detection technology show that coherent upconverison of quantum states is feasible, where the photon statistics would be conserved consequently [10,11]. Therefore, it is possible to count the visible replicas via frequency upconversion to realize PNRDs around 1 μm.

Based on coincidence frequency upconversion presented in Section 3, photon-number-resolving detection at 1.04 μm has been realized. The experimental setup was schematically illustrated in Fig. 8, consisting of two parts: synchronization fiber laser system for the pump pulses and signal photons, and photon-number-resolved detection system based on coincidence frequency upconversion and the Si-MPPC. The signal photons at 1.04 μm were then upconverted by the synchronized pump pulses at 1.55 μm in the PPLN crystal. Then the SFG photons passed through a group of filters before impinging on the Si-MPPC with a multimode fiber pigtail. The Si-MPPC (Hamamatsu Photonics S10362-11-100U) was composed of 10×10 APDs which were arranged on an effective active area of 1 mm$^2$ with a photon detection efficiency of 16.0% (including the fiber coupling efficiency). When the incident photons were injected onto different pixels, the output voltage of the superposition from all APD pixels was proportional to the number of incident photons [40]. The histogram of the peak voltage from the output of Si-MPPC was showed in Fig. 9(a). The pulse area spectrum featured a series of peaks representing the different photon number states. By fitting each peak to Gaussian function, the probability distribution was obtained as shown in Fig. 9(b). The area under each Gaussian curve gave the number of events presenting that photon number state. The area of each peak could be normalized by the total area to give the probability distribution. As the input light was in coherent state, the upconverted photons statistics obeyed the Poissionian law. By fitting the experimental data according to Poisson distribution, we got the detected photon number of 4.65. With different incident photon flux, the photon numbers detected by SPCM were shown with circles in Fig. 9(c). Since SPCM could not discriminate more than one photon per shot, it was obviously saturated with large incident photon numbers. In contrast, as shown with squares in Fig. 9(c), the detected photon numbers of Si-MPPC linearly increased with the incident photon numbers. It showed that the Si-MPPC could correctly identify the photon numbers per pulse with a large dynamic range, which was promising in few-photon-level detection.

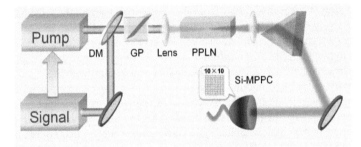

**Figure 8.** Experimental setup for the 1.04 μm photon-number-resolved detection. DM, dichroic mirror; GP, Glan prism; Si-MPPC, silicon multipixel photon counter.

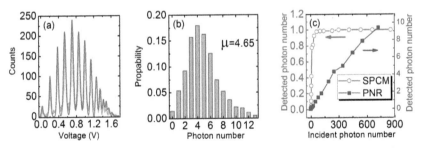

**Figure 9.** (Color online) (a) Output voltage amplitude histogram for the upconverted photons. (b) Photon number distribution. (c) Detected photon numbers by SPCM (circles) and Si-MPPC (squares) as a function of incident photon numbers.

The photon-number-resolving performance was improved by reducing the background counts with a synchronous pump as the coincidence gate and reducing the intrinsic parametric fluorescence influence with long-wavelength pumping. As a result, a total detection efficiency of 3.7% was achieved with a quite low noise probability per pulse of 0.0002. Such a low background noise probability could remarkably improve the sensitivity of the frequency-upconversion PNRD. The remarkable decrease in background noise would optimize applications such as quantum entanglement and quantum teleportation [42] and improve the signal-to-noise ratio of widely used light detection and ranging system, since the background counts would randomly couple into the modes of the quantum states, which may significantly affect the original photon number distribution [38]. The approach may find promising applications in various quantum optical experiments using nonclassical light sources to demonstrate the features of quantum states around 1 μm [43].

## 4.2. Ultrasensitive infrared imaging

Realization of ultra-sensitive infrared imaging has critical importance for applications such as astronomy, medical diagnosis, night-vision technology and chemical sensing. Currently, the infrared imaging detectors are available, like the commonly used linear InGaAs photodiode array. However, suffering from the severe dark current, the sensitivity of such detectors is largely limited [44]. Even though liquid-nitrogen-cooling can provide a solution for a smaller dark current, the additional cryogenic cooling device reduces the feasibility for lots of applications. Unlike in the infrared regime, imaging in the visible regime can be readily implemented by silicon charged coupled devices (CCDs) with high resolution and high efficiency. Moreover, recent progresses in sensor technology have led to the development of electron multiplying CCDs (EMCCDs) which is capable of single-photon detection. By leveraging the high sensitivity of EMCCDs, ultra-sensitive infrared imaging can be expected with nonlinear frequency upconversion of the infrared electric field to the visible spectral region [10,11]. Here we will introduce a few-photon-level two-dimensional infrared imaging detector by coincidence frequency upconversion. The upconversion imaging apparatus is shown in Fig. 10.

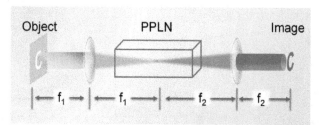

**Figure 10.** Schematic for infrared imaging by frequency upconversion.

In our experiment, the pump and signal sources were taken from two fiber lasers mode-locked at 19.1 MHz [33]. The signal beam illuminated a transmission mask with a character "C" to form the object beam shown in Fig. 11 (a). By a dichroic mirror, the object beam and the pump beam were then combined into a 4-f imaging system with lens L1 ($f_1$=250 mm) and L2 ($f_2$=300 mm). Fourier plane was arranged right at the middle of 50-mm-long PPLN crysta. In order to facilitate the type I quasi-phase matching of the PPLN crystal, a Glan prism was employed before the crystal for enforcing the polarization [45]. The temperature of the crystal was optimized at 104.3 °C for the grating period of 11.0 μm. The upconverted image at 622 nm was captured by a silicon EMCCD (Andor iXon3 897) bearing 512×512 pixels. The pixel size was 16×16 μm, which was very suitable for high spatial resolution imaging. To improve the signal-to-noise ratio, the EMCCD was thermoelectrically cooled to -85 °C.

With the pump power of 40.0 mW, corresponding to a peak power of about 210 W, the internal conversion efficiency of 27% could be inferred by correcting for the filtering transmittance. The object beam was attenuated to 2.0 photons per pulse. The upconverted imaging photons were then registered by the EMCCD with integration time of 30 s and accumulated for 50 times. The upconverted image was shown in Fig. 11 (c), from which the character "C" could be identified. Thanks to the pulsed pump field together with the long-wavelength pumping, the background noise was remarkably reduced. The theoretical calculation based on coherent imaging theory was given in Fig. 11 (b), which was in good agreement with the experimental result. The image blurring was attributed to the spatial filtering relating to the point spread function in the upconversion imaging system [7]. To improve the resolution, large pump beam profile at the Fourier plane would be better to enwrap the transformed object field as much as possible. However, increasing the diameter of the pump beam would decrease the intensity, thus reduce the conversion efficiency. Thus optimization of the trade-off should be in consideration.

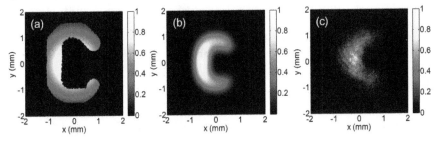

**Figure 11.** (a) The coherently illuminated mask. The theoretical (b) and measured (c) upconverted images at the image plane.

In summary, we demonstrated full 2D infrared image upconversion at few-photon level with a high conversion efficiency of 31%. The infrared image at 1040 nm was upconverted to the visible regime where the imaging photons were registered by the silicon EMCCD with high sensitivity and resolution. The imaging performance was remarkably improved with the reduction of the background noise by the synchronized pulsed excitation of the 1549 nm pump source, as well as the long-wavelength pumping [46]. Such all-optical upconversion imaging technique can offer an attractive method for ultra-sensitive infrared imaging.

## 5. Conclusion

In this review chapter, the quantum theory of the frequency upconversion is introduced and indicates that either in single-mode regime and multi-mode regime, complete quantum transduction can be realized in principle with two necessary requirement, sufficiently intensive pump field and perfect phase matching. Several upconversion systems with high conversion efficiency are presented and discussed in detail. Synchronous pumping frequency upconversion system shows superior performance with high conversion efficiency and low background noise. Thanks to the short time window of the synchronized pump pulse together with long-wavelength pumping scheme, the detection sensitivity was improved remarkably by reducing the background noise. This technique facilitates not only many traditional applications, such as classical optical communication, imaging, photobiology, and astronomy, but also novel quantum optics applications, such as quantum interface to transfer quantum entanglement, linear optical quantum gates, single photon polarization switches, and nonlinear control of single photons.

## Author details

Kun Huang, E Wu, Xiaorong Gu, Haifeng Pan and Heping Zeng

*State Key Laboratory of Precision Spectroscopy, East China Normal University, China*

## Acknowledgement

This work was funded in part by National Natural Science Fund of China (10990101, 60907043, 61127014 & 91021014), International Cooperation Projects from Ministry of Science and Technology (2010DFA04410), Key project sponsored by the National Education Ministry of China (109069), Research Fund for the Doctoral Program of Higher Education of China (20090076120024), ECNU Reward for Excellent Doctors in Academics.

## 6. References

[1]  Boyd R (2008) Nonlinear Optics, Academic Press, ISBN 978-0123694706, New York.

[2]  Huang J, Kumar P (1992) Observation of Quantum Frequency Conversion. Phys. Rev. Lett. 68: 2153–2156.

[3]  Albota M, Wong F (2004) Efficient Single-Photon Counting at 1.55 μm by Means of Frequency Upconversion. Opt. Lett. 29: 1449–1451.

[4]  Pan H, Dong H, Zeng H, Lu W (2006) Efficient Single-Photon Counting at 1.55 μm by Intracavity Frequency Upconversion in a Unidirectional Ring Laser. Appl. Phys. Lett. 89: 191108.

[5]  Vandevender A, Kwiat P (2004) High Efficiency Single Photon Detection via Frequency Up-Conversion. J. Mod. Opt. 51: 1433–1445.

[6]  VanDevender A, Kwiat P (2007) Quantum Transduction via Frequency Upconversion. J. Opt. Soc. Am. B, 24: 295-299.

[7]  Pedersen C, Karamehmedović E, Dam J, Tidemand-Lichtenberg P (2009) Enhanced 2D-Image Upconversion Using Solid-State Lasers. Opt. Express 17: 20885-20890.

[8]  Kuzucu O, Wong F, Kurimura S, Tovstonog S (2008) Time-Resolved Single-Photon Detection by Femtosecond Upconversion. Opt. Lett. 33: 2257-2259.

[9]  Zhang Q, Langrock C, Fejer M, Yamamoto Y (2008) Waveguide-Based Single-Pixel Up-Conversion Infrared Spectrometer. Opt. Express 16: 19557–19561.

[10] Tanzilli S, Tittel W, Halder M, Alibart O, Baldi P, Gisin N, Zbinden H (2005) A Photonic Quantum Information Interface. Nature 437: 116–120.

[11] Rakher M, Ma L, Slattery O, Tang X, Srinivasan K (2010) Quantum Transduction of Telecommunications-Band Single Photons from a Quantum Dot by Frequency Upconversion. Nature Photonics 4: 786–791.

[12] Xu H, Ma L, Mink A, Hershman B, Tang X (2007) 1310-nm Quantum Key Distribution System with Up-Conversion Pump Wavelength at 1550 nm. Opt. Express 15: 7247–7260.

[13] Fejer M, Magel G, Jundt D, Byer R (1992) Quasi-Phase-Matched Second Harmonic Generation: Tuning and Tolerances. IEEE J. Quantum Electron. 28: 2631-2654.

[14] Takesue H, Diamanti E, Langrock C, Fejer M, Yamamoto Y (2006) 1.5 μm Single Photon Counting Using Polarization-Independent Up-conversion Detector. Opt. Express 26: 13067-13072.

[15] Ma L, Slattery O, Tang X (2010) NIR Single Photon Detectors with Up-Conversion Technology and Its Applications in Quantum Communication Systems. InTech. pp. 315–336.

[16] Ramelow S, Fedrizzi A, Poppe A, Langford N, Zeilinger A (2012) Polarization-Entanglement-Conserving Frequency Conversion of Photons. Phys. Rev. A 85: 013845.

[17] Gu X, Huang K, Pan H, Wu E, Zeng H (2012) Photon Correlation in Single-Photon Frequency Upconversion. Opt. Express 20: 2399-2407.

[18] Huang K, Gu X, Ren M, Jian Y, Pan H, Wu G, Wu E, Zeng H (2011) Photon-Number-Resolving Detection at 1.04 μm via Coincidence Frequency Upconversion. Opt. Lett. 36: 1722-1724.

[19] Midwinter J, Warner J (1967) Up-conversion of Near Infrared to Visible Radiation in Lithium-Meta-Niobate, J. Appl. Phys. 38: 519-523.

[20] Kim Y, Kulik S, Shih Y (2001) Quantum Teleportation of A Polarization State with a Complete Bell State Measurement. Phys. Rev. Lett. 86: 1370-1373.

[21] Hum D, Fejer M (2007) Quasi-Phasematching. C. R. Physique 8: 180–198.

[22] Pan H, Zeng H (2006) Efficient and Stable Single-Photon Counting at 1.55 μm by Intracavity Frequency Upconversion. Opt. Lett. 31: 793-795.

[23] Langrock C, Diamanti E, Roussev R, Yamamoto Y, Fejer M, Takesue H (2005) Highly Efficient Single-Photon Detection at Communication Wavelengths by Use of Upconversion in Reverse-Proton-Exchanged Periodically Poled LiNbO₃ Waveguides. Opt. Lett. 30: 1725-1727.

[24] Dong H, Pan H, Li Y, Wu E, Zeng H (2008) Efficient Single-Photon Frequency Upconversion at 1.06 μm with Ultralow Background Counts. Appl. Phys. Lett. 93: 071101.

[25] Gu X, Li Y, Pan H, Wu E, Zeng H (2009) High-Speed Single-Photon Frequency Upconversion with Synchronous Pump Pulses. IEEE J. Sel. Top. Quantum Electron. 15: 1748-1752.

[26] Gu X, Huang K, Li Y, Pan H, Wu E, Zeng H (2010) Temporal and Spectral Control of Single-Photon Frequency Upconversion for Pulsed Radiation. Appl. Phys. Lett. 96: 131111.

[27] Huang K, Gu X, Pan H, Wu E, Zeng H (2012) Synchronized Fiber Lasers for Efficient Coincidence Single-Photon Frequency Upconversion. IEEE J. Sel. Top. Quantum Electron. 18: 562-566.

[28] Pan H, Wu E, Dong H, Zeng H (2008) Single-Photon Frequency Up-Conversion with Multimode Pumping. Phys. Rev. A 77: 033815.

[29] Rakher M, Ma L, Davanco M, Slattery O, Tang X, Srinivasan K (2011) Simultaneous Wavelength Translation and Amplitude Modulation of Single Photons from a Quantum Dot. Phys. Rev. Lett. 107: 083602.

[30] Eckstein A, Brecht B, Silberhorn C (2011) A Quantum Pulse Gate Based on Spectrally Engineered Sum Frequency Generation. Opt. Express 19: 13770-13778.

[31] Pomarico E, Sanguinetti B, Thew R, Zbinden H (2010) Room Temperature Photon Number Resolving Detector for Infrared Wavelengths. Opt. Express 18: 10750-10759.

[32] Albota M, Wong F, Shapiro J (2006) Polarization-Independent Frequency Conversion for Quantum Optical Communication. J. Opt. Soc. Am. B 23: 918-924.

[33] Rusu M, Herda R, Okhotnikov O (2004) Passively Synchronized Two-Color Mode-Locked Fiber System Based on Master-Slave Lasers Geometry. Opt. Express 12: 4719-4724.

[34] Hao Q, Li W, Zeng H (2009) High-Power Yb-Doped Fiber Amplification Synchronized with a Few-Cycle Ti: Sapphire Laser. Opt. Express 17: 5815-5821.

[35] Li Y, Gu X, Yan M, Wu E, Zeng H (2009) Square Nanosecond Mode-Locked Er-Fiber Laser Synchronized to a Picosecond Yb-Fiber Laser. Opt. Express 17: 4526-4532.

[36] Xu P, Ji S, Zhu S, Yu X, Sun J, Wang H, He J, Zhu Y, Ming N (2004) Conical Second Harmonic Generation in a Two-Dimensional $\chi^{(2)}$ Photonic Crystal: A Hexagonally Poled LiTaO₃ Crystal. Phys. Rev. Lett. 93: 133904.

[37] Allevi A, Bondani M, Andreoni A (2010) Photon-Number Correlations by Photon-Number Resolving Detectors. Opt. Lett. 35: 1707-1709.

[38] Waks E, Diamanti E, Sanders B, Bartlett S, Yamamoto Y (2004) Direct Observation of Nonclassical Photon Statistics in Parametric Down-Conversion. Phys. Rev. Lett. 92: 113602.

[39] Afec I, Natan A, Ambar O, Silberberg Y (2009) Quantum State Measurements Using Multipixel Photon Detectors. Phys. Rev. A 79: 043830.

[40] Eraerds P, Legré M, Rochas A, Zbinden H, Gisin N (2007) SiPM for Fast Photon-Counting and Multiphoton Detection. Opt. Express 15: 14539-14549.

[41] Wu G, Jian Y, Wu E, Zeng H (2009) Photon-Number-Resolving Detection Based on InGaAs/InP Avalanche Photodiode in the Sub-Saturated Mode. Opt. Express 17: 18782-18787.

[42] Honjo T, Takesue H, Kamada H, Nishida Y, Tadanaga O, Asobe M, Inoue K (2007) Long-Distance Distribution of Time-Bin Entangled Photon Pairs over 100 km Using Frequency Up-Conversion Detectors. Opt. Express 15: 13957-13964.

[43] Vasilyev M, Choi S, Kumar P, D'Ariano G (2000) Tomographic Measurement of Joint Photon Statistics of the Twin-Beam Quantum State. Phys. Rev. Lett. 84: 2354–2357.

[44] Liang Y, Jian Y, Chen X, Wu G, Wu E, Zeng H (2011) Room-Temperature Single-Photon Detector Based on InGaAs/InP Avalanche Photodiode with Multichannel Counting Ability. IEEE Photon. Tech. Lett. 23: 115–117.

[45] Thew R, Zbinden H, Gisin N (2008) Tunable Upconversion Photon Detector. Appl. Phys. Lett. 93: 071104.

[46] Pelc J, Zhang Q, Phillips C, Yu L, Yamamoto Y, Fejer M (2012) Cascaded Frequency Upconversion for High-Speed Single-Photon Detection at 1550 nm. Opt. Lett. 37: 476-478.

# Chapter 6

# Single-Molecule Recognition and Dynamics with Pulsed Laser Excitation

Guofeng Zhang, Ruiyun Chen, Yan Gao, Liantuan Xiao and Suotang Jia

Additional information is available at the end of the chapter

## 1. Introduction

Single-molecule spectroscopy has evolved from a specialized variety of optical spectroscopy into a versatile tool used to address a broad range of questions in physics, chemistry, biology, and materials science. Due to the ultra short time duration, the pulsed laser can be applied widely in the research of single-molecule dynamics. In the chapter, we will discuss the laser pulse application in two aspects of single-molecule detection and spectroscopy: fast recognition of single molecules, and manipulation of interfacial electron transfer dynamics.

The chapter is organized as follows. In the part of fast recognition of single molecules, we will discuss the Mandel's Q-parameter of single-event photon statistics for single-molecule fluorescence, recognition of single molecules using Q parameter, and the influence of signal-to-background ratio and the error estimates for fast recognition of single molecules. In the other part of the chapter, we will discuss the manipulation of interfacial electron transfer dynamics. First, we will introduce the principle of fluorescence lifetime measurement. Next, we will show the experiment results of the single-molecule and the ensemble under the external electric currents. Last, we will present our analysis and discussion for the results.

## 2. Fast recognition of single molecules

Although people most often think about and model molecule systems in terms of individuals, experimental science has been dominated by measurements that result in ensemble averages. This has traditionally hidden much of the rich variety present at microscopic scales. Although detecting single molecules optically was an old dream, the first convincing detection of a single molecule was achieved in 1989 by Moerner and Kador in an absorption measurement (Moerner et al., 1989). Optical spectroscopy offers a wealth of information on the structure, interaction, and dynamics of molecule species. Soon here after, fluorescence was shown to provide a much better signal-to-noise ratio, in cryogenic condition (Orrit et al., 1990) as well as at room temperature (Shera et al., 1990; Basché et al., 1992). The microscopy of single molecules at room temperature took off in 1993 with Betzig and Chichester's detection of immobilized molecules on a solid surface by means of excitation with a near-field optical source (Betzig et al., 1993). The scope of the method expanded suddenly when several groups (Nie et al., 1994; Trautman et al., 1994; Funatsu et al., 1995; Dickson, et al., 1996) showed that single molecules could be detected at ambient conditions with a simple confocal microscope. Single-molecule microscopy by fluorescence at room temperature has now become a versatile and general technique, opening investigations of the nanoworld (Orrit, 2002; Bartko et al., 2002). In above experiments it is critical to ascertain that the observed signal actually comes from a single molecule, but not the random mixture of emissions from nearby molecules. Usually low concentration and equivalently small excitation volumes are the most common experimental strategies (Deniz et al., 1999).

Once experimental conditions favoring single-molecule detection are satisfied, a number of criteria have to be met to ascertain that the observed signal actually comes from a single emitter (Michalet et al., 2002 ; Nie et al., 1994). The most common criteria is to detect an antibunching curve by two-time correlation measurements: based on quantum properties of single photon states, the absence of coincidence at zero delay gives clear evidence of single photon emission (Brunel et al., 1999). For several molecules, coincident emission of photons by the different molecules are likely and will result in an autocorrelation function that does not cancel out for zero time-delay.

Although the phenomenon of photon antibunching is demonstrated most clearly by two-time correlation measurements, it is, in principle, also exhibited by the probability $P(n)$ that $n$ photons are emitted (or detected) in a given time interval $T$. Antibunching implies sub-Poissonian statistics, in the sense that the probability distribution is narrower than a Poisson distribution with the same $\langle n \rangle$ (Mandel, 1979). Traditional photon number counting statistics (Huang et al., 2006) is seriously affected by the blinking in the fluorescence, due to the molecular triplet state. Then single event photon statistics (Huang et al., 2007; Treussart et al., 2002) is suggested to character the single-molecule fluorescence. And based on moment analysis, Mandel's $Q$-parameter, which is defined in terms of the first two photon count moments, is an attractive alternative to two-time correlation measurements. It is quite robust with respect to molecular triplet state effects (Sanchez-Andres et al., 2005). In this work we suggest a novel approach to distinguish single-molecule system based on single event photon statistics characterization of single-molecule fluorescence. Using Hanbury Brown and Twiss (HBT) configuration (Hanbury Brown et al., 1956), by analyzing and comparing the Mandel parameters of actual single molecule fluorescence and ideal double molecule fluorescence, we present a new criterion based on single event photon statistics measurement.

## 2.1. Detection system of single-molecule fluorescence

Standard confocal microscopy technique (Michalet & Weiss, 2002) which is the commonly used experimental setup for single-molecule fluorescence detection is summarized in Fig.

1(a) , the emitted light from molecules is focused on a pinhole in order to reject out-of-focus background light and then recollimated onto two single photon counting modules (SPCM) after partition by a 50/50 beamsplitter (BS), which is a standard HBT configuration. For each detected photon, the SPCM would generate a TTL pulse.

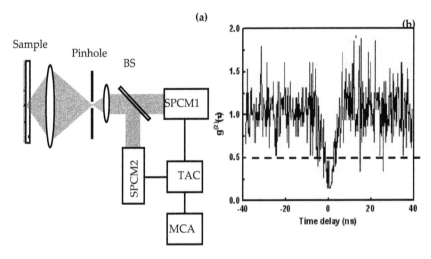

**Figure 1.** (a) Schematic of a confocal setup used for single-molecule fluorescence detection. (b) The second-order autocorrelation function measured, which indicates that fluorescence of a single molecule was being detected.

## 2.2. Detection of second-order coherence function

The signals from SPCMs are inputed to a time-to-amplitude converter (TAC). The start-stop technique with allows us to build a coincidences histogram as a function of time delay between two consecutive photodetections on each side of the beamsplitter. Then the pulses from TAC, whose amplitude is proportional to the time delay, are discriminated according to their heights and accumulated by a multichannel pulse height analyzer (MCA). The second-order degree of coherence at zero delay $g^{(2)}(0)$ can be directly gained and displayed on the screen, as shown in Fig. 1(b).

In experiment the effect of background signals makes it impossible for the perfect absence of coincidence at zero delay. Commonly as long as the second-order degree of coherence $g^{(2)}(\tau)$ at zero delay ($\tau=0$) is less than 0.5, which corresponds with two ideal molecules detected, the fluorescence can be considered from a single emitter. Obviously if $g^{(2)}(0)$ is bigger than 0.5, it indicates that the fluorescence detected is emitted from more than one molecule. In contrast, in Fig. 1(b), $g^{(2)}(0)$ is smaller than 0.5, which indicates that one single molecule was being detected. This method is widely used to distinguish single-molecule in experiments. However, for a small number of blinking molecules, the total number of coincident emission might be too low to reject the single-molecule hypothesis with the antibunching curve. Furthermore the criteria need long time (typically several minutes to tens minutes, depending on the mean photon number $\langle n \rangle$ and the dead-time of detection system) to detect enough photons to accumulate and display a visible antibunching curve. It is severely affected by the molecule's photostability. And the existence of irreversible photobleaching compelled the long time for single molecules recognition to be insufferable in experiments. Using this method, the fluorescent photons detected should be very weak ($\langle n \rangle$ is less than 0.1), otherwise the start-stop technique will bring an error that can not be ignored.

(a)                                                                    (b)

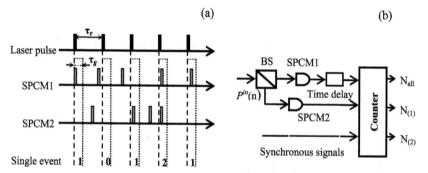

**Figure 2.** (a) The single event photon statistics measurement. The sample gate time $\tau_g$, dead time for SPCM $\tau_d$ and laser pulse period $\tau_r$ fulfill $\tau_g < \tau_d < \tau_r$. (b) The schematic of detection setup used for single event photon statistics measurement. The synchronous signals provide a counting time-gate.

## 2.3. Single event photon statistics characterization of single-molecule fluorescence

The single event photon statistics measurement is described in Fig. 2(a). In order to eliminate the background signals as much as possible, time-gated technique (Shera, et al., 1990) is used after the excitation pulse. The records within the time gates are considered to be the detected signals while all records outside the time gates are rejected. This time-filtering procedure can filter out the real photodetection events from the most of non-synchronous background photocounts not rejected by optical filters, and is often an efficient way to improve the signal-to-noise ratio. The time gate duration must be shorter than the laser period and much longer than the molecule excited-state lifetime so that the probability of discarding a fluorescence signal is negligible. In the above experiments the gate durations are usually ten times the radiative lifetime of the molecule (Alléaume et al., 2004). Furthermore most after-pulse which will show up within 3 ns after the detector dead time, can be eliminated by time-gated technique, when time gate is shorter than the dead time. For the influence of detectors' dead time, each SPCM gives only one count (supposedly with 100% detection efficiency) for one or more than one incident photon within the dead time. For every excitation pulse cycle, the number of detected photons cannot exceed two if we using two identical SPCMs operating in the photon counting regime compulsory for our system. The existence of detectors' dead time in each detection channels will results in a non-linear relation-ship between detected photon statistics and source photon statistics. Thus the measured photon probabilities should be corrected.

Fig. 2(b) shows the schematic of our detection setup used for single event photon statistics measurement. The synchronous signals provide a counting time-gate and a nanosecond time delay box is used to compensate the different transport distance between two detection channels. The numbers of pulse cycles which are detected $N_{all}$, in which only one photon is detected $N_{(1)}$, and in which two photons are detected $N_{(2)}$ can be directly gained from a counter. Then the ratio of $N_{(1)}$ to $N_{all}$ and $N_{(2)}$ to $N_{all}$ respectively is the probability of detected one-photon and of detected two-photon in every pulse cycle.

Denoting by $P^{in}(n)$ the photon number probability distribution of incoming light on HBT detection set-up, the non-linear transformation relating this probability to the detected photon probability $P(n = 0, 1, 2)$ is simply calculated for 'ideal' detectors. The 'ideal' means that each SPCM clicks with 100% efficiency immediately upon receiving a photon, but that no more than one click can occur in a given repetition period. The joint probability of detecting $i$ photons on SPCM1 and $j$ photons on SPCM2 can be written as $P(i, j)$, $i, j = 0$ or 1. Actually, there are total four measured photon probabilities: $P(0, 0)$, $P(0, 1)$, $P(1, 0)$ and $P(1, 1)$.

With our experimental detection scheme, random splitting of photons on two sides of 50/50 beamsplitter gives

$$P(0) = P(0,0) = P^{in}(0),$$

$$P(1) = P(0,1) + P(1,0) = \sum_{n\geq 1}^{\infty} P^{in}(n)\frac{1}{2^{n-1}}, \tag{1}$$

$$P(2) = P(1,1) = \sum_{n\geq 2}^{\infty} P^{in}(n)\left(1 - \frac{1}{2^{n-1}}\right).$$

The mean photon number per excitation pulse period is

$$\langle n \rangle = P(1) + 2P(2). \tag{2}$$

## 2.4. Mandel's parameter for single event photon statistics

In order to quantify the fluctuations of the number of photons detected per pulse, an important figure of merit is the Mandel parameter $Q = (\langle(\Delta n)^2\rangle - \langle n\rangle)/\langle n\rangle$, where $\langle n \rangle$ is the average number of photons detected within a time interval $T$ and $\langle(\Delta n)^2\rangle$ is the mean variance (Short et al., 1983). The parameter $Q$ is a natural measure of the departure of the variance of the photon number $\langle n \rangle$ from the variance of a Poisson process, for which $Q = 0$. Negative and positive $Q$-values indicate sub- and super-Poissonian behavior, respectively.

From single event photon statistics probability the Mandel parameter can be computed directly

$$Q = \frac{2P(2)}{\langle n \rangle} - \langle n \rangle. \tag{3}$$

Note that an ideal single molecule should produce photons containing exactly one photon per pulse, P(2)=0, would yield $Q_i = -\langle n \rangle =, -\eta$, which means Q is only limited by the detection efficiency. Negative Q confirms that single-molecule fluorescence indeed exhibits sub-Poissonian photon statistics which is an explicit feature of a quantum field.

## 2.5. Distinguishing single-molecule using Q-parameter

The Mandel's Q-parameter provides an alternative for differentiating single-molecule fluorescence system. The Q-parameter of ideal double molecules (which is used to model two molecules that can not be separated by diluting) fluorescence can be used as a boundary between that of an actual single molecule fluorescence and actual double molecules fluorescence. From this boundary we can deduce an explicit criterion of single molecules based on photon statistics.

Now suppose that in the excitation volume there has been more than one molecule, the number is s, and all of them are ideal single photon emitters. We call $\eta$ the overall detection efficiency, which includes the optical collection efficiency, all linear propagation losses and quantum efficiency of photon detectors. Then the photon number probability distribution of incoming light on detection set-up is

$$P_I^{in}(n) = \frac{s!}{n!(s-n)!}(1-\eta)^{s-n}\eta^n. \tag{4}$$

Applying Eq. (4) in Eq. (1) one can show

$$P(0) = (1-\eta)^s,$$

$$P(1) = \sum_{n \geq 1}^{s} \frac{s!}{n!(s-n)!}(1-\eta)^{s-n}\eta^n \frac{1}{2^{n-1}}, \tag{5}$$

$$P(2) = \sum_{n \geq 2}^{s} \frac{s!}{n!(s-n)!}(1-\eta)^{s-n}\eta^n \left(1 - \frac{1}{2^{n-1}}\right).$$

Then

$$\langle n \rangle = s\eta(1-\eta)^{s-1} + \sum_{n \geq 2}^{s} \frac{2^n-1}{2^{n-1}} \frac{s!}{n!(s-n)!}(1-\eta)^{s-n}\eta^n. \tag{6}$$

Now suppose that in the excitation volume there are two ideal molecules excited and the fluorescent photons emitted are detected without background noise at all. Denote by $Q_D$ and $P_D(n)$ the Mandel parameter and single event photon statistics probability of the fluorescence signals. From Eq. (4) one obtains

$$P_D(0) = (1-\eta)^2,$$

$$P_D(1) = 2\eta - \frac{3}{2}\eta^2, \tag{7}$$

$$P_D(2) = \frac{1}{2}\eta^2,$$

$$\langle n \rangle = 2\eta - \frac{1}{2}\eta^2.$$

In an experiment a problem we have to face is the background signals, which includes radiation from the environment and the dark counts of the SPCM. Because the scattering background light from surroundings can be thought as a thermal field with very large bandwidth and very short coherent time, the usual photon counting time (nano-seconds) discussed here is much longer than the coherent time and the photocounts of such time-average stationary background show a Poisson distribution. In a dark environment, the SPCM also generates random dark counts that follow a Poisson distribution. Both of these two random counts appear in the Poisson distribution, and thus we can use a weak coherent field with a Poisson photon distribution $P_B^{in}(n) = e^{-\eta\gamma}(\eta\gamma)^n/n!$ with $\gamma$ to simulate the backgrounds. Actual single-molecule fluorescence can be modeled as the superposition of an ideal single molecule and a background emission that can be modeled as a Poisson distribution. Denote by $Q_A$ and $P_A(n)$ the Mandel parameter and single event photon probability statistics of the actual single molecules fluorescence. Using Eq. (1) and Eq. (4), the photon statistics probability can be written as

$$P_A(0) = e^{-\eta\gamma}(1-\eta),$$

$$P_A(1) = 2(e^{-\eta\gamma/2} - e^{-\eta\gamma}) + \eta(2e^{-\eta\gamma} - e^{-\eta\gamma/2}), \tag{8}$$

$$P_A(2) = (1 - e^{-\eta\gamma/2})^2 + \eta(e^{-\eta\gamma/2} - e^{-\eta\gamma}).$$

Then

$$\langle n \rangle = 2(1 - e^{-\eta\gamma/2}) + \eta e^{-\eta\gamma/2}.$$

The measured mean photon number of fluorescence signals is $S = \eta$, and the measured mean photon number of background signals is $B = 2(1 - e^{-\eta\gamma/2})$. According to Eq. (7), we get

$$P_A(0) = (1-S)\left(1 - \frac{B}{2}\right)^2,$$

$$P_A(1) = (S + B - SB)\left(1 - \frac{B}{2}\right),$$ (9)

$$P_A(2) = \frac{BS}{2} + \frac{B^2}{4} - \frac{B^2 S}{4},$$

$$\langle n \rangle = B + S(1 - \frac{B}{2}).$$

When

$$P_A(1) \geq 2\sqrt{P_A(2)} - 3P_A(2),$$

the signal-to- background ratio (SBR) can be expressed as

$$SBR = \frac{S}{B} = \frac{P_A^{\,2}(1)}{2P_A(2)}.$$ (10)

This equation can be directly applied for the measurement of SBR in experiment.

Using Eq. (6) one can show that with the same average photon number $\langle n \rangle$, if $P_A(2) < P_D(2)$, then $Q_A < Q_D$. From Eq. (10) one can obtains

$$P_A(2) < \frac{1}{2}\left(2 - \sqrt{4 - 2\langle n \rangle}\right)^2.$$ (11)

At the same time,

$$P_A(1) > \langle n \rangle - \left(2 - \sqrt{4 - 2\langle n \rangle}\right)^2.$$ (12)

It is not possible to have fluorescence from more than one fluorophore with it's $Q$-parameter smaller than $Q_D$. So when Eq. (11) or (12) is satisfied, $Q_A < Q_D$, the fluorescence can be deduced origin from a single molecule system.

## 2.6. Signal-to-background ratio effect on the criterion

However low background signal is an important precondition of the criterion of Eqs. (11) and (12). With the high background signals, $Q_A > Q_D$ will come into existence for a single molecule system. This criterion will not be applicable. So it is necessary to make certain the range of SBR when Eqs. (11) and (12) is satisfied.

Corresponding to Eqs. (11) and (12) using Eqs. (8) and (9) one can obtains

$$SBR_A > SBR_0 = \frac{\sqrt{\langle n \rangle^2 - 2\left(2 - \sqrt{4 - 2\langle n \rangle}\right)^2}}{\langle n \rangle - \sqrt{\langle n \rangle^2 - 2\left(2 - \sqrt{4 - 2\langle n \rangle}\right)^2} - \frac{1}{2}\left(\langle n \rangle - \sqrt{\langle n \rangle^2 - 2\left(2 - \sqrt{4 - 2\langle n \rangle}\right)^2}\right)^2}. \quad (13)$$

It is well known that SBR of the actual single molecule system varies with average photon number $\langle n \rangle$. Based on Eq. (13), the SBR curve versus $\langle n \rangle$ is shown in Fig. 3. When $\langle n \rangle$ is between 0 and 1, the variation range of SBR is within 1.63 to 2.41. So with the same $\langle n \rangle$ when $SBR_A > SBR_0$, for actual single molecule fluorescence and ideal double molecules fluorescence, it has $Q_A < Q_D$. Especially when SBR > 2.41, the Eqs. (11) and (12) can be the criterion used to distinguish single molecule system. Contrarily when $SBR_A < SBR_0$, even an actual single molecule system can not satisfy the Eqs. (11) and (12). It is obvious that SBR is also an important parameter of distinguishing single molecule system. Nevertheless most of the SBR in single molecule fluorescence experiments is large than 2.41.

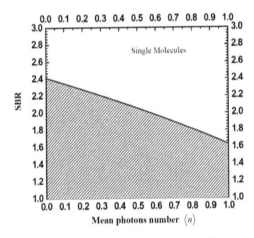

**Figure 3.** The curve of SBR as a function of the mean photon number $\langle n \rangle$ for actual single molecular photon source. The range out of the shaded portion means the ones that can use the criterion of Eqs. (11) and (12).

## 2.7. Results and discussion

After error correction from appendix one can obtain a more sufficient criterion,

$$P_A(1) > \langle n \rangle - \left(2 - \sqrt{4 - 2\langle n \rangle}\right)^2 - \delta P_A(1) + \langle \Delta P_A(1) \rangle = P_1, \quad (14)$$

$$P_A(2) < \frac{1}{2}\left(2 - \sqrt{4 - 2\langle n \rangle}\right)^2 - \delta P_A(2) - \langle \Delta P_A(2) \rangle = P_2, \quad (15)$$

In the above inequalities the critical values $P_1$, $P_2$ are impacted by three parameters, the mean photon number $\gamma$ of Poissonian background, the mean overall detection efficiency $\eta$, and factor $\Delta$ which represents the unbalance of two channels of imperfect detection system.

Now let us assume that the size of sample cycles $M$ was $10^4$. In this way the statistical fluctuation of the $\langle \Delta P_A \rangle$ induced by finite sample pulse cycles detected can be negligible, which is less than 1/100 of the statistics probability itself. The error $\delta P_A$ caused by imperfect detection system is the main correction factor of the critical value.

Fig. 4 shows the critical values $P_1$, $P_2$ as the function of mean overall detection efficiency $\eta$ and factor $\Delta = 0.3$, $\gamma = 0.2$, which corresponds to SBR = 5. It is found that the critical values $P_1$, $P_2$ all increase with increasing efficiency $\eta$ as respected, however the $P_1$ increases slowly while $P_2$ increases fast. Furthermore because in single molecule experiments $P_A(2)$ is less than $P_A(1)$, the effect of the error on $P_2$ is more evident than on $P_1$. So using $P_A(1)$ as the criterion is more feasible.

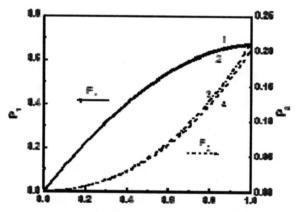

**Figure 4.** The critical values $P_1$ (solid line), $P_2$ (dashed line) as a function of mean overall detection efficiency $\eta$. And factor $\Delta = 0.3$, $\gamma = 0.2$, which corresponds to SBR = 5. $P_1$ (solid line) using left coordinate, $P_2$ (dashed line) using right coordinate. Curves 2 and 3 are the critical values $P_1$, $P_2$ without error estimate. Curves 1 and 4 are the critical values $P_1$, $P_2$ after error correction.

The experiment data, shown in Table I, is obtained by using typical single molecule experiment setup (Basché et al., 1992). The sample were dye molecules Cy5 ($5 \times 10^{-10}$ M, Molecular Probes) doped in Polymethyl Methacrylate (PMMA, Sigma-Aldrich) polymer films. Molecules in the sample were excited by 50 ps pulses at a wavelength of 635 nm and repetition rate of 2 MHz, generated by a ps pulsed diode laser (PicoQuant, PDL808), at the focus of a confocal inverted microscope (Nikon). Fluorescence photons were detected in two detection channels, using two identical single photon counting modules (SPCMs, PerkinElmer SPCM-AQR-15).

For example, in a typical experiment to sample 1, during 299613 periods (about 149 ms) there are 13917 recorded photons including 13902 single photon events, 15 two-photons events. These data allow us to extract the photon probabilities P(0) = 0.9535, P(1) = 0.0464, and P(2) = 5 × 10$^{-5}$ and the mean number of detected photon per pulse $\langle n \rangle$ = 0.0465, then yield Mandel parameter Q = - 0.04435.

We observed fluorescence photons for three sample molecules in one polymer film.

As shown in Table 1, for sample 1 and sample 2, both P(1) are much more than critical values P$_1$ which indicates that a single molecule was being detected; for sample 3, P(1) is less than critical values P$_1$ which indicates that more than one molecule was being detected. As a reference, the interrelated results for coherent light (pulse laser) are attached both from theoretical analysis and experimental measurement. The typical time for single event photon statistics measurement is about 150 ms.

| | $n$ | P(1) | P$_1$ | g$^{(2)}$ (0) | SM? | SBR |
|---|---|---|---|---|---|---|
| Sample 1 | 0.0465 | 0.0464 | 0.04590 | 0.19 | Yes | 21 |
| Sample 2 | 0.0372 | 0.0370 | 0.03685 | 0.30 | Yes | 7 |
| Sample 3 | 0.0521 | 0.0508 | 0.05150 | 0.65 | No | -- |
| Coherent light (theory) | 0.1046 | 0.0991 | 0.0991 | 1.0 | -- | -- |
| Coherent light (experiment) | 0.1046 | 0.0992 | 0.0995 | 0.98 | -- | -- |

**Table 1.** Experimental results of single event photon statistics for Cy5 molecules and coherent light. The mean number $\langle n \rangle$ of detected photons per pulse is calculated by the measured values P(1) and P(2) from Eq. (2). The critical value P$_1$ is estimated from Eq. (14), the relative error is less than 1%. And the g$^{(2)}$ (0) is measured by two-time correlation measurements. Whether the sample detected is a single molecule is determined by that P(1) is more than critical value P$_1$ or not. And the determination is confirmed by the measured value g$^{(2)}$ (0). To a single molecule, the SBR is acquired from Eq. (9).

## 3. Manipulation of interfacial electron transfer dynamics

Interfacial electron transfer (IET) dynamics play an important role in many chemical and biological processes (Weiss, 1676; Wang et al., 2009). However, IET processes are usually very complex due to high dependence on its local environments. Single-molecule spectroscopy has led to many surprises and has now become a standard technique to study complex structures (Kulzer et al., 2010) or dynamics (Orrit et al., 2002; Uji-i et al., 2006; Zhang et al., 2010) including photoinduced, excited-state intramolecular, and IET dynamics. The optical signals of single molecules provide information about dynamics of their nanoscale environment, free from space and time averaging. Single-molecule studies of photoinduced electron transfers in the enzyme flavin reductase have revealed formational fluctuation at multiple time scales (Yang et al., 2003). Single-molecule studies of photosensitized electron transfers on dye containing nanoparticles also showed fluorescence fluctuations and blinking, and the fluctuation dynamics were found to be inhomogeneous from molecule to molecule and from time to time (Funatsu et al., 1995). Recently developed single-molecule spectroelectrochemistry extends single-molecule approaches to ground state IET by simultaneously modulating the electrochemical potential while detecting single molecule fluorescence (Palacios et al., 2006; Lei et al., 2009).

Indium tin oxide (ITO) films are the most widely used material as a transparent electrode of organic light emitting diode and also in other devices like solar cells (Friend et al., 1999; Tak et al., 2002; Hanson et al., 2005). It is interesting to study electron transfer of ITO to suitably modify interactions at the interfaces of dissimilar materials so that desired electronic properties of devices incorporating them can be realized. Single dye molecules dispersed on the semiconductor surface of ITO were used to measure IET from excited cresyl violet molecules to the conduction band of ITO or energetically accessible surface electronic states under ambient conditions by using a far-field fluorescence microscope, and single-molecule exhibited a single-exponential electron transfer kinetics (Lu et al., 1997). Here we apply an external electric current (EEC) to manipulate the IET rate between single dye molecules and its neighboring ITO nanoparticles by probing the fluorescence intensity change of individually immobilized single dye molecules dispersed in ITO film.

## 3.1. Experimental section

Cover glass substrates were cleaned by acetone, soap solution, milliQ water sonication and irradiation with ultraviolet lamp. The single molecules samples were prepared by spin coating (3000 rpm) a solution of 1,1'- dioctadecyl - 3, 3, 3′, 3′-tetramethylindodi-carbocyanine (DiD, $10^{-9}$M, Molecular Probes) in chlorobenzene onto the cover glass substrate. The chemical structure of DiD molecule is shown in Fig. 5(a). The indium tin oxide (ITO) was purchased from Sigma-Aldrich (Product Number: 700460, dispersion, <100 nm particle size (DLS), 30 wt. % in isopropanol, composition: $In_2O_3$ 90%, $SnO_2$ 10%). ITO film in hundreds of nanometer thicknesses was spin-coated onto the dye molecules. Two aluminum leads were fixed to the ITO film, and the interval between the two leads is about 4 mm. After vacuum-dried, the samples were further covered with a poly-(methyl methacrylate) (PMMA) (Mw = 15,000, Tg = 82 °C, Aldrich) film in order to insulate oxygen. The samples were subsequently annealed in vacuum at 350 K for 5 h to remove residual solvent, oxygen and to relax influences of the spin coating technique on the polymer conformations. The method is depicted schematically in Fig. 5(b).

A 70 picosecond pulse diode laser of $\lambda = 635$ nm (PicoQuant, PDL808) with a repetition rate of 40 MHz was used to excite single dye molecules. The output of the pulse laser passed through a polarizing beamsplitter cubes (New Focus 5811) to obtain linear polarization light. A 1/4 wave-plate was used to change the polarized laser into circular polarization light. The laser beam was sent into a conventional inverted fluorescence microscope (Nikon ECLIPSE TE2000-U) from its back side, reflected by a dichroic mirror (BrightLine, Semrock, Di01-R635-25x36), and focused by an oil immersion objective lens (Nikon, 100 ×, 1.3 NA) onto the upper sample surface of the cover glass substrate. The fluorescence was collected by the same objective lens and then passed through the dichroic mirror, an emission filter (BrightLine, Semrock, FF01-642/LP-25-D), and a notch filter (BrightLine, Semrock, NF03-633E-25), is focused onto a 100 μm pinhole for spatial filtering to reject out-of-focus photons.

Fluorescence photons were subsequently focused through a lens and collected by a single-photon detector (PerkinElmer, SPCM-AQR-15). A piezo-scan stage (Piezosystem jena, Tritor 200/20 SG) with an active x-y-z feedback loop mounted on the inversion microscope was used to scan the sample over the focus of the excitation spot, producing a two-dimensional fluorescence imaging. All measurements were conducted in a dark compartment at room temperature. We use an alternative power source to supply EEC for the DID/ITO film system. Here, the applied electric current is proportional to the amplitude of applied external bias voltage, with the relation being $I = ZU$ with $Z = 0.87 ohm^{-1}$. The dye molecules embedded in the ITO film are distinct from the molecules in electric field experiments (Hania et al 2006). In the electric field experiments, a large electric field intensity of about 100 V/μm is needed, and dyes must be insulated from the electrode. In our experiment, dye molecules directly contact with ITO nanoparticles for IET controlling with much low voltage needed (less than 0.02 V/μm). In order to distinguish with the normal electric field experiments, we discuss the results ensuing from EECs here.

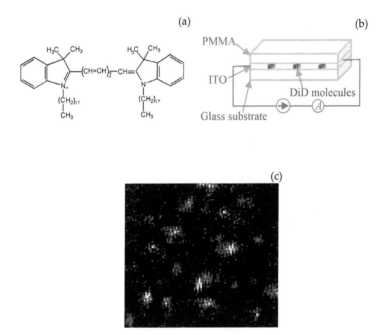

**Figure 5.** Fig. 5 (a) Structure of the 1, 1'-dioctadecyl-3, 3, 3', 3'-tetramethylindodicarbocyanine (DiD) dyes molecule. (b) Schematic structure of the sample preparation. (c) Confocally scanned fluorescence image (10 μm × 10 μm) of DiD molecules dispersed in an indium tin oxide (ITO) film. Each bright feature may be attributed to a single DiD molecule.

## 3.2. Results and discussion

### 3.2.1. Fluorescence imaging of single dye molecules in ITO film

Fluorescence imaging using a confocal arrangement has superior sensitivity for spectroscopic measurements and is most suitable for studying single-molecule behavior in dilute sample. Fig. 5(c) displays the confocal fluorescence image of the single DiD molecules within a 10 μm × 10 μm area, which is obtained by scanning a sample containing randomly placed isolated single fluorescent molecules dispersed in an ITO film. The imaging is taken in 225 s (150 pixels × 150 pixels), with a pixel integration time of 10 ms. The single molecules are excited with laser excitation intensity of 1.5 kW/cm$^2$ at 40 MHz. The concentration of the DiD molecules was kept at such a low level that either one molecule or no molecules were in the focus. The bright features in the image then represent the fluorescence from individual molecules. The full width at half maximum of typical spots is about 300 nm. Meanwhile, it is also found that the spots have different intensities due to the molecular orientations and local environments. We have found that the average intensities of the molecules were about an order of magnitude weaker than that of DiD molecules in PMMA without ITO under the same experimental conditions. The weaker signal intensities result from the molecules undergoing the IET with surrounding ITO nanoparticles (Lu, et al., 1997).

### 3.2.2. Electric current response of the ensemble fluorescence

To study the IET dynamics between dye molecules and ITO nanoparticles, we placed dyes embedded in ITO film. An EEC was added to the DiD/ITO film to manipulate the IET rate. The molecules fluorescence intensity was measured by centering ensemble or one molecule in the laser focus and recording the transient fluorescence intensity, while a time-dependent EEC was applied.

We measured the EEC response of the ensemble at different dyes concentration, and the ensemble averaged study would establish how the average fluorescence intensity and lifetime are affected by the electrical current. The Fig. 6 shows three typical ensemble average results. Fig. 6(a) shows that the fluorescence intensity of an ensemble decreases rapidly while EEC applyed. Fig. 6(b) shows a fluorescence quenching trajectory of an ensemble under the EEC. However, fluorescence can be enhanced sometimes for some ensembles at relative small EECs as shown in Fig. 6(c). The similar fluorescence enhancement was also observed(Palacios et al., 2009), which was explained that potential-induced modulation of the excited state reduction processes (i.e., electron transfer from ITO to the molecules) dominate the low-potential fluorescence-modulation effect. We can find from the Fig. 6 that the ensembles do not show complete fluorescence quenching at the relative large EEC. This may be due to that those dye molecules in the ensemble are not in good contact with the ITO nanoparticles, which shows poor IET.

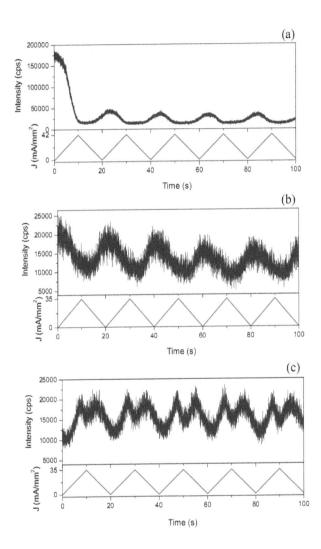

**Figure 6.** Three typical patterns of ensemble fluorescence intensity trajectories that were obtained while repeatedly applying a triangle wave EEC sequence to ITO shown by the bottom red curves. (a) The fluorescence intensity shows a fast quenching while applying the EEC to the ITO. (b) The fluorescence intensity shows a decrease while applying the EEC to the ITO. (c) The trajectory shows that fluorescence increases at smaller EEC and then decreases at larger EEC.

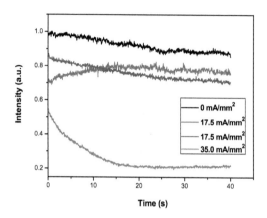

**Figure 7.** Normalized fluorescence intensity trajectories of 25 ensembles obtained at different EECs. The black curve shows the average of 25 ensemble trajectories obtained at zero current; the blue curve shows the average of 16 ensemble trajectories with fluorescence quenching obtained at the currentF of 17.5 mA/mm²; the red curve shows the average of 9 ensemble trajectories with fluorescence enhancement obtained at the current density of 17.5 mA/mm²; the green curve shows the average of 25 ensemble trajectories obtained at the current density of 35.0 mA/mm².

Normalized fluorescence intensity trajectories of 25 ensembles with different dyes concentration obtained under different EECs is shown in Fig. 7. Under relative small EECs, some ensembles show fluorescence quenching and the other ensembles show fluorescence enhancement, and the red curve and blue curve are constructed by sorting trajectories with fluorescence enhancement and fluorescence quenching at the current density of 17.5 mA/mm². In our experiment, for approximately 30% of the ensemble data show fluorescence enhancement effects at relative small current. The green curve is the average of 25 ensemble trajectories obtained at the current density of 35.0 mA/mm², and it shows fluorescence quenching of the ensemble at a large current. The various responses to EECs arise from the heterogeneity of site-specific molecules.

### 3.2.3. Electric current response of single-molecule fluorescence

We have detected several hundreds of single DiD molecules and all of the molecules sensitively responded to EEC. Fig. 8 shows the typical fluorescence emission for different individual DiD molecules in dependence of the EEC as a function of time. Fig. 8(a) is the fluorescence intensity time trace of one DiD molecule as an EEC of 32.0 mA/mm² periodically applied to the ITO film, which shows that the EEC can effectively quench the fluorescence emission of single-molecule. Fluorescence blinking observed in the trace shows a single DiD molecule emission and the blinking is related to the triplet state and/or charge transfer between single molecule and its local environment. While the electric current

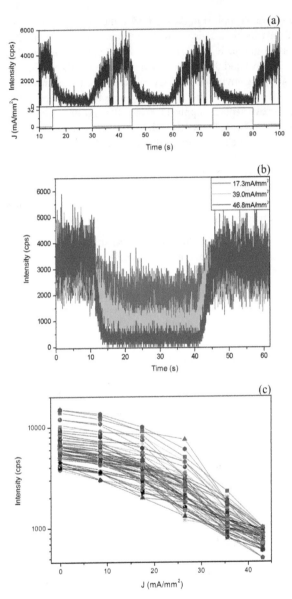

**Figure 8.** EEC response of DiD single-molecule emission intensity. (a) One single molecule fluorescence emission patterns as electric current applied periodically to the electrodes. (b) Another single-molecule emission intensities under different electric currents applied between 10 s and 40 s. (c) Single-molecule fluorescence intensity trajectories (data points) for 75 DiD single molecules that were obtained while applying different EECs.

density of 32.0 mA/mm$^2$ is applied to the electrodes, the fluorescence intensities of the single molecule exhibit an exponential decay with the time constant about 2.24 ± 0.23 s. The fluorescence emission gradually recovers to the initial value which needs about 10 s after switching off the EEC. We find that both of the decay time and the recovering time depend on the molecular neighboring environment. The emission intensities of another single molecule at different electric currents are showed in Fig. 8(b), note that the rates of intensity decay are different, when electric currents are applied between 10 s and 40 s. These intensity decay traces were fitted by single-exponential function with the time constants of 5.80 ± 1.20 s, 3.20 ± 0.54 s and 1.38 ± 0.09 s under the electric current density of 17.3 mA/mm$^2$, 39.0 mA/mm$^2$, and 46.8 mA/cm$^2$ respectively. Note that the bigger current, the faster response. The intensities recover to the initial values which need about 7.5 s after switching off the EEC. It is also found from the Fig. 8(b) that the 46.8 mA/mm$^2$ current density (and even bigger) can quench almost completely the fluorescence emission. The two molecules have the different recovering times, which arises from the heterogeneity of site-specific molecular interactions.

In addition, we record the average fluorescence intensities of each single DiD molecule at different currents. The statistical data of 75 DiD single molecules are shown in Fig. 8(c). All the DiD single molecules exhibit intensity decrease with the large EEC.

We present a model, as shown in the Fig. 9, to explain the results of fluorescence quenching of single-molecule. The schematic representation of energy levels, basic photoinduced and electric current driving processes, and multiple electron transfer cycles in DiD / ITO system are shown.

When an EEC is applied to the ITO film, the Fermi level of the ITO is tuned by the potential. With a positive potential, the Fermi level of ITO is decreased and there is more driving force for the forward electric transfer (FET) and the electron transfer of ground state (GET), but backward electron transfer (BET) is suppressed simultaneously.

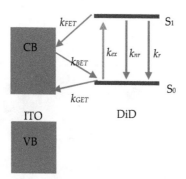

**Figure 9.** Schematic representation of energy levels, basic photoinduced and electric current driving process, and multiple electron transfer cycles in DiD / ITO system. $k_r$, radiative decay rate; $k_{nr}$, intrinsic nonradiative decay rate; $k_{FET}$, the rate of forward interfacial electron transfer from excited molecule to semiconductor; $k_{BET}$, the rate of backward electron transfer; $k_{GET}$, the rate of electron transfer from ground state of molecule to ITO; CB, the conduction band of ITO; VB, the valence band of ITO.

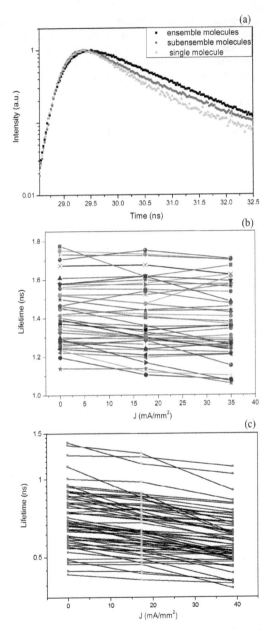

**Figure 10.** (a) Fluorescence decays for an ensemble molecules (black curve), a subensemble molecules (red curve) and a single molecule (green curve) in ITO film, measured by time correlated photon counting. The decay curves are fitted with single-exponential decay with the time constant about 1.41

ns, 1.16 ns and 0.90 ns respectively. (b) The fluorescence lifetimes of ensemble average for 45 ensembles at three different electric currents. The average fluorescence lifetime is 1.43 ns at 0 mA/mm$^2$, 1.40 ns at 17.5 mA/mm$^2$, and 1.38 ns at 35.0 mA/mm$^2$ respectively. (c) The fluorescence lifetimes of 65 single molecules at three different electric currents. The average fluorescence lifetime is 0.73 ns at 0 mA/mm$^2$, 0.67 ns at 17.3 mA/mm$^2$, and 0.59 ns at 39.0 mA/mm$^2$ respectively.

While the BET is suppressed completely at a large EEC, the fluorescence is completely quenched. When EEC is switched off, the Fermi level of ITO will gradually recover to its original value and the driving force for the FET and GET will decrease, simultaneously the BET will increase, thus the fluorescence will recover gradually.

### 3.2.4. Electric current dependence of fluorescence lifetime

Lu and Xie have presented that each single molecule exhibits a single-exponential IET dynamics in dye molecules / ITO film system, the rate of FET, $k_{FET}$, can be measured by the fluorescence decay of excited dye molecules. And the changes of fluorescence lifetimes were attributed to the FET (Guo et al., 2007; Wang et al., 2009; Jin, 2010).

Fig. 10(a) shows three typical fluorescence decay curves for an ensemble (average intensity is about 800 k cps), a subensemble (average intensity is about 40k cps) and a single DiD molecule (average intensity is about 8 k cps ) in ITO film. The decay curves are fitted with single-exponential decay with the time constant about 1.41 ns, 1.16 ns and 0.90 ns respectively. Accordingly, the lifetimes of DiD single-molecule are shorter than the 2.5 ns ~ 3.0 ns lifetime measured in the polymer (Vallée et al., 2003). Fig. 10(b) shows that the fluorescence lifetimes of ensemble average for 45 ensembles at three different electric currents, ranging from 1.1 ns to 1.8 ns. The average fluorescence lifetime is 1.43 ns at 0 mA/mm$^2$, 1.40 ns at 17.5 mA/mm$^2$, and 1.38 ns at 35.0 mA/mm$^2$ respectively. Fig. 10(c) shows fluorescence lifetimes of 65 DiD single molecules at three different electric currents, ranging from ~300 ps to 1.4 ns. The average fluorescence lifetime is 0.73 ns at 0 mA/mm$^2$, 0.67 ns at 17.3 mA/mm$^2$, and 0.59 ns at 39.0 mA/mm$^2$ respectively. Unfortunately, we cannot measure a single-molecule lifetime shorter than 300 ps due to the limited sensitivity for time resolution of the instrumental response. As mentioned above, the FET rate is highly sensitive to the interactions between the dyes and ITO nanoparticles. When some dye molecules in the ensemble that are not in good contact with the ITO nanoparticles there will induce the poor IET dynamics. Thus, fluorescence lifetime of the ensemble is longer than the lifetime of single-molecule. In the typical dye-photosensitization system, FET can shorten the fluorescence lifetime and reduce fluorescence quantum yield of the dye molecules. Under the EEC, the small change of fluorescence lifetimes may indicate the change of the FET rate, and the fluorescence quenching may be mainly dominated by the BET and GET.

## 4. Conclusion

We present a fast and robust method to recognise single molecules based on single event photon statistics. Mandel's Q-parameter provides an attractive approach to two-time correlation measurements, because it is easy to implement, requires little time, and is immune with respect to the effects of molecular triplet state. Compared with common two-time correlation measurements, our approach has some advantages: (1) the effect of molecular triplet state can be ignored, whereas its effect can only be contained in the non-perfect detection efficiency analysis; (2) ~ ms level measurement time is needed as only ~$10^4$ fluorescence photons are needed for photon statistic; (3) it is not limited only to weak photons emitted, which means it is independent with the fluorophores photon intensity. The method can also be applied for the other single emitters recognition, such as single atoms, quantum dots and color centers.

Individual DiD dye molecules were dispersed in ITO semiconductor films as probes of IET dynamics, which should help to understand the intrinsic properties of electron transfer at interface between organic molecules and transparent semiconductor materials. While the EEC was used to drive the FET and GET, and suppress the BET, the change of $k_{FET}$ induces the change of fluorescence lifetime and the increasing $k_{GET}$ and decreasing $k_{BET}$ would quench the fluorescence. Due to the inhomogeneous nature of the interactions from molecule to molecule, the lifetime under EEC is inhomogeneous. The EEC dependence of lifetime distribution clearly demonstrates a manipulated IET dynamics, which can be revealed by the single-molecule experiments instead of by the ensemble-averaged measurements. The results could open up a new path to manipulate single-molecule electron transfer dynamics by using EEC while measuring single-molecule fluorescence intensity and lifetime simultaneously.

## Author details

Guofeng Zhang, Ruiyun Chen, Yan Gao, Liantuan Xiao and Suotang Jia
*State Key Laboratory of Quantum Optics and Quantum Optics Devices,*
*Laser Spectroscopy Laboratory, Shanxi University, Taiyuan, China*

## Acknowledgement

The project sponsored by 973 Program (Nos. 2012CB921603, 2010CB923103), 863program (No. 2011AA010801),the Natural Science Foundation of China (Nos. 11174187 ˒ 11184145, 60978018 and 10934004), NSFC Project for Excellent Research Team (No. 61121064), the Natural Science Foundation of Shanxi province, China (No. 2011091016), TSTIT and TYMIT of Shanxi, and Shanxi Province Foundation for Returned scholars.

## 5. References

Alléaume, R.; Treussart, F.; Courty, J. M. & Roch, J. F. (2004). Experimental open-air quantum key distribution with a single-photon source, *New J. Phys.*, Vol. 6, No. 92, (March 2004) page number (1-12), ISSN: 1367-2630

Bartko, A. P.; Xu, K. & Dickson, R. M. (2002). Three-dimensional single molecule rotational diffusion in glassy state polymer films, *Phys. Rev. Lett.*, Vol. 89, No. 2, (July 2001) page numbers (026101-1-4), ISSN: 0031-9007

Basché, T. & Moerner, W. E. (1992). Energetics of negatively curved graphitic carbon, *Nature*, Vol. 355, No. 6358, (January 1992) page numbers (333-335), ISSN: 0028-0836

Betzig, E. & Chichester, R. J. (1993). Single molecules observed by near-field scanning optical microscopy, *Science*, Vol. 262, No. 5138, (September 1993) page number (1422-1425), ISSN: 0036-8075

Brunel, C.; Lounis, B.; Tamarat P. & Orrit, M. (1999). Triggered source of single photons based on controlled single molecule fluorescence, *Phys. Rev. Lett.*, Vol. 83, No. 14, (February 1999) page numbers (2722-2725), ISSN: 0031-9007

Deniz, A. A.; Dahan, M.; Grunwell, J. R.; Ha, T.; Faulhaber, A. E.; Chemla, D. S.; Weiss, S. & Schultz, P. G. (1999). Single-pair fluorescence resonance energy transfer on freely diffusing molecules: Observation of Förster distance dependence and subpopulations, *Proc. Nat. Acad. Sci.USA*, Vol. 96, (March 1999) page number (3670-3675), ISSN: 0027-8424

Dickson, R. M.; Norris, D. J.; Tzeng, Y. L. & Moerner, W. E. (1996). Three-dimensional imaging of single molecules solvated in pores of poly(acrylamide) gels, *Science*, Vol. 274, No. 5289, (November 1996) page number (966-968), ISSN: 0036-8075

Friend, R. H.; Gymer, R. W.; Holmes, A. B.; Burroughes, J. H.; Marks, R. N.; Taliani, C.; Bradely, D. D. C.; Dos Santos, D. A.; Bredas, J. L.; Logdlund, M. & Salaneck, W. R. (1999). Electroluminescence in conjugated polymers, *Nature*, Vol. 397, (January 1999) page numbers (121-128), ISSN: 0028-0836

Funatsu, T.; Harada, Y.; Tokunaga, M.; Saito, K. & Yanagida, T. (1995). Imaging of single fluorescent molecules and individual ATP turnovers by single myosin molecules in aqueous solution, *Nature*, Vol. 374, (April 1995) page numbers (555-559), ISSN: 0028-0836

Guo, L.; Wang, Y. & Lu, P. H. (2010). Combined single-molecule photon-stamping spectroscopy and femtosecond transient absorption spectroscopy studies of interfacial electron transfer dynamics, *J. Am. Chem. Soc.*, Vol. 132, No. 6, (January 2010) page numbers (1999-2004), ISSN: 0002-7863

Hanbury, B. R. & Twiss, R. Q. (1956). Correlation between photons in two coherent beams of light, *Nature*, Vol. 177, (January 1956) page numbers (27-29), ISSN: 0028-0836

Hania, P. R., Thomsson, D. & Scheblykin, I. G. (2006). Host matrix dependent fluorescence intensity modulation by an electric field in single conjugated polymer chains, *J. Phys. Chem. B*, Vol. 110, No. 51, (December 2006) page numbers (25895-25900), ISSN: 1520-6106

Hanson, E. L.; Guo, J.; Koch, N.; Schwartz, J. & Bernasek, S. L. (2005). Advanced surface modification of indium tin oxide for improved charge injection in organic devices, *J. Am. Chem. Soc.*, Vol. 127, No. 28, (June 2005) page numbers (10058-10062), ISSN: 0002-7863

Huang, T.; Dong, S. L.; Guo, X. j.; Xiao, L. T. & Jia, S. T. (2006). Signal-to-noise ratio improvement of photon counting using wavelength modulation spectroscopy, *Appl.*

*Phys. Lett.*, Vol. 89, No. 6, (November 2005) page number (061102-061102-3), ISSN: 0003-6951

Huang, T.; Wang, X. B.; Shao, J. H.; Guo, X. J.; Xiao, L. T. & Jia, S. T. (2007). Single event photon statistics characterization of a single photon source in an imperfect detection system, *J. Lumin.*, Vol. 124, No. 2 (August 2005) page number (286-290), ISSN: 0022-2313

Jin, S.; Snoeberger III, R. C.; Issac, A.; Stockwell, D.; Batista, V. S. & Lian, T. (2010). Single-molecule interfacial electron transfer in donor-bridge-nanoparticle acceptor complexes, *J. Phys. Chem. B*, Vol. 114, No. 45, (March 2010) page numbers (14309-14319), ISSN: 1520-6106

Kulzer, F.; Xia, T. & Orrit, M. (2010). Single molecules as optical nanoprobes for soft and complex matter, *Angew. Chem. Int. Ed.*, Vol. 49, No. 5, (January 2010) page numbers (854-866), ISSN: 1433-7851

Lei, C. H.; Hu, D. H. & Ackerman, E. (2009). Clay nanoparticle supported single-molecule fluorescence spectroelectrochemistry, *Nano Lett.*, Vol. 9, No. 2, (January 2009) page numbers (655-658), ISSN: 1530-6954

Lu, H. P. & Xie, X. S. (1997). Single-molecule kinetics of interfacial electron transfer, *J. Phys. Chem. B*, Vol. 101, No. 15, (February 1997) page numbers (2753-2757), ISSN: 1520-6106

Mandel, L. (1979). Sub-poissonian photon statistics in resonance fluorescence, *Opt. Lett.*, Vol. 4, No. 7, (March 1979) page number (205-207), ISSN: 0146-9592

Michalet, X. & Weiss, S. (2002). Single-molecule spectroscopy and microscopy, *C. R. Physique*, Vol. 3, No. 5, (March 2002) page number(619-644), ISSN: 1631-0705

Moerner, W. E. & Kador, L. (1989). Optical detection and spectroscopy of single molecules in a solid, *Phys. Rev. Lett.*, Vol. 62, No. 21, (March 1989) page numbers (2535-2538), ISSN: 0031-9007

Nie, S.; Chiu, D. T. & Zare, R. N. (1994). Probing individual molecules with confocal fluorescence microscopy, *Science*, Vol. 266, No. 5187, (November 1994) page number (1018-1021), ISSN: 0036-8075

Orrit, M. (2002). Single-molecule spectroscopy: The road ahead, *J. Chem. Phys.*, Vol. 119, No. 24, (December 2002) page numbers (10938-10946), ISSN: 0021-9606

Orrit, M. & Bernard, J. (1990). Single pentacene molecules detected by fluorescence excitation in a p-terphenyl crystal. *Phys. Rev. Lett.*, Vol. 65, No. 21, (July 1990) page numbers (2716-2719), ISSN: 0031-9007

Palacios, R. E.; Fan, F. R. F.; Bard, A. J. & Barbara, P. F. (2006). Single-molecule spectroelectrochemistry, *J. Am. Chem. Soc.*, Vol. 128, No. 28, (June 2006) page numbers (9028-9029), ISSN: 0002-7863

Palacios, R. E.; Chang, W. S.; Grey, J. K.; Chang, Y. L.; Miller, W. L.; Lu, C. Y.; Henkelman, G.; Zepeda, D.; Ferraris, J. & Barbara, P. F. (2009). Detailed single-molecule spectroelectrochemical studies of the oxidation of conjugated polymers, *J. Phys. Chem. B*, Vol. 113, No. 44, (October 2009) page numbers (14619-14628), ISSN: 1520-6106

Sanchez-Andres, A.; Chen, Y. & Müller, J. D. (2005). Molecular brightness determined from a generalized form of mandel's Q-parameter, *Biophys. J.*, Vol. 89, No. 5, (May 2005) page number (3531-3547), ISSN: 0006-3495

Shera, E. B.; Seizinger, N. K.; Davis L. M.; Keller, R. A. & Soper, S. A. (1990). Detection of single fluorescent molecules, *Chem. Phys. Lett.*, Vol. 174, No.6, (July 1990) page numbers (553-557), ISSN: 0009-2614

Short, R. & Mandel, L. (1983). Observation of sub-poissonian photon statistics, *Phys. Rev. Lett.*, Vol. 51, No. 5, (April 1983) page numbers (384-387), ISSN: 0031-9007

Tak, Y. H.; Kim, K. B.; Park, H. G.; Lee, K. H. & Lee, J. R. (2002). Criteria for ITO (indium–tin-oxide) thin film as the bottom electrode of an organic light emitting diode, *Thin Solid Films*, Vol. 411, No. 1, (May 2002) page numbers (12-16), ISSN: 0040-6090

Trautman, J. K.; Macklin, J. J.; Brus, L. E. & Betzig, E. (1994). Imaging and time-resolved spectroscopy of single molecules at an interface, *Science*, Vol. 272, No. 5259, (April 1996) page number (255-258), ISSN: 0036-8075

Treussart, F. ; Alléaume, R. ; Le Floc'h, V., Xiao, L. T.; Courty, J. M. ; & Roch, J. F. (2002). Direct measurement of the photon statistics of a triggered single photon source, *Phys. Rev. Lett.*, Vol. 89, No. 9, (February 2002) page numbers (093601-1-4), ISSN: 0031-9007

Uji-i, H.; Melnikov, S. M.; Deres, A.; Bergamini, G.; De Schryver, F.; Herrmann, A.; Müllen, K.; Enderlein, J. & Hofkens, J. (2006). Visualizing spatial and temporal heterogeneity of single molecule rotational diffusion in a glassy polymer by defocused wide-field imaging, *Polymer*, Vol. 47, No.7, (March 2006) page numbers (2511-2518), ISSN: 0032-3861

Vallée, R. A. L.; Tomczak, N.; Kuipers, L.; Vancso, G. J. & van Hulst, N. F. (2003). Single molecule lifetime fluctuations reveal segmental dynamics in polymers, *Phys. Rev. Lett.*, Vol. 91, No. 3, (July 2003) page numbers (1038301-1--1038301-4), ISSN: 0031-9007

Wang, Y. M.; Wang, X. F.; Ghosh, S. K. & Lu, H. P. (2009). Probing single-molecule interfacial electron transfer dynamics of porphyrin on TiO2 nanoparticles. *J. Am. Chem. Soc.*, Vol. 131, No. 4, (January 2009) page numbers (1479-1487), ISSN: 0002-7863

Wang, Y. M.; Wang, X. F. & Lu, H. P. (2009). Probing single-molecule interfacial geminate electron–cation recombination dynamics, *J. Am. Chem. Soc.*, Vol. 131, No. 25, (June 2009) page numbers (9020-9025), ISSN: 0002-7863

Weiss, S. (1999). Fluorescence spectroscopy of single biomolecules, *Science*, Vol. 283, No. 5408, (March 1999) page numbers (1676-1683), ISSN: 0036-8075

Yang, H.; Luo, G. B.; Karnchanaphanurach, P.; Louie, T. M.; Rech, I.; Cova, S.; Xun, L. Y. & Xie, X. S. (2003). Protein conformational dynamics probed by single-molecule electron transfer, *Science*, Vol. 302, No. 5643, (October 2003) page numbers (262-266), ISSN: 0036-8075

Zhang, G. F.; Xiao, L. T.; Zhang, F.; Wang, X. B. & Jia, S. T. (2010). Single molecules reorientation reveals the dynamics of polymer glasses surface, *Phys. Chem. Chem. Phys.*, Vol. 12, (December 2009) page numbers (2308-2312), ISSN: 1463-9076

# Chapter 7

# Generation of Tunable THz Pulses

J. Degert, S. Vidal, M. Tondusson, C. D'Amico, J. Oberlé and É. Freysz

Additional information is available at the end of the chapter

## 1. Introduction

Many phenomena in materials science, physics, chemistry, biology and medicine involve fundamental processes with a spectral signature in the THz range [20]. However, for many years, these processes have been little studied due to the lack of high quality THz sources. Fortunately, the recent developments in THz technology have filled this so-called THz gap [28]. With the emergence of reliable THz sources, the need for arbitrarily shaped THz pulses dedicated to specific applications like communications, signal processing, spectroscopy or coherent/optimal control is more and more felt. Concerning coherent control, selective excitations in the THz spectral region of phonon modes in molecular crystal [41], charge oscillations in semiconductor heterostructures [5, 26] or phonon-polaritons in ferroelectric crystal [9, 40] have already been achieved by means of temporally shaped near-infrared laser pulses. From this point of view, the possibilities of control would be increased if one makes it possible to excite directly the system with shaped THz pulses [27]. To this end, several methods have been developed to extend the generation of arbitrary pulse shapes from the visible and mid-infrared spectral region up to the THz range [1, 8, 10, 16–19, 21, 30, 31, 43]. All are based on the generation of THz pulses by excitation of different types of emitters with spatially and/or temporally shaped optical pulses.

Here, we will focus our attention on the generation, tuning and shaping of narrow-band THz pulses. Such pulses have already been implemented by means of various pulse shaping techniques which differ from each other by their respective spectral width, their ease of use and their versatility. Some of them use traditional pulse shaping setup: for example, Sohn et al. [30] used a specially designed mask placed in the Fourier plane of a zero-dispersion line to produce modulated optical pulses subsequently photomixed to generate THz pulses tunable from 0.5 to 3 THz, with a spectral width of $\sim$ 500 GHz. Despite its tunability, this method is limited by the use of a non programmable mask. Likewise, using optical pulse shaping with a liquid crystal spatial light Fourier filter combined with optical rectification in ZnTe, Ahn et al. were able to generate THz waves tunable from 0.5 to 2 THz, with a spectral bandwidth of $\sim$ 200 GHz [1]. To do this, they have implemented a Gerchberg-Saxton algorithm to find the best spectral phase required for the optical pulse in order to generate the desired THz pulse. However, this technique, characterized by its great ability to generate various pulse shapes, has one drawback: it relies on the ability of the algorithm to converge towards the expected solution. On the other hand, interesting examples of tunable narrow-band THz pulses with possibilities of shaping the waveform have also been reported in lithium niobate crystals. For example, Lee et al. [16, 18, 19] employed a single pulse or a pair of temporally separated optical pulses rectified in PPLN crystals with a specially engineered domain structure, generating very narrow THz pulses (bandwidth $\sim$ 25 GHz) tunable between 0.5 and 2.5 THz. Nevertheless, it requires a new crystal with a specific domain structure for each temporal shape or an adjustment of the crystal position to tune the frequency. At last, using a transient polarization grating induced by two femtosecond laser pulses propagating in a LiNbO$_3$ crystal, Stepanov et al. [31] were able to generate and to shape THz pulses tunable from 0.5 to 3 THz with a bandwidth of 100 GHz. In this experiment, the tunability is controlled by the angle between the pump beams and the shaping is accomplished by filtering with a slit or a shield the spatial intensity distribution of the pump beams. While the spatial filtering can be done with a liquid crystal modulator, enabling thus to do optimal control experiments, the control of the tunability is less flexible.

In the present chapter, we present two different techniques that make it possible to generate tunable THz pulses. The first one, that can be regarded as analytical, is based on the spectral tailoring of an ultrashort femtosecond laser pulse, whereas the second one is relying on the spatial shaping of the transverse profile of the femtosecond laser beam. In both techniques, liquid crystal devices are used to modulate in the spatial or spectral domain the femtosecond pulses that are incident on a nonlinear crystal in which optical rectification is taking place. Hence, it is flexible and needs no moving mechanical parts or Mach-Zender like interferometers. We demonstrate that the first one makes it possible to generate THz pulses tunable between 0.5 and 2.5 THz with spectral bandwidth as narrow as 140 GHz. The second one, based on a geometrically-assisted optical rectification technique, enables us to generate THz spatiotemporal interferences in the intermediate field zone beyond the rectifying crystal. We will show that the spatiotemporal properties of the generated THz field in the intermediate zone are tightly tied to the geometry of the transverse profile of the laser beam attaining the rectifying crystal. Therefore by shaping the transverse beam profile one can change the temporal and the transverse profile of the emitted THz pulse.

## 2. Model and preliminary analysis of THz generation by optical rectification in zinc blende crystals

Whatever the way used to obtain shaped THz waves, it is strongly correlated to the physical mechanisms involved in the THz wave generation. Here, since our attention will be focused on THz generation by optical rectification in Zinc Blende crystals, we are going to remind, at first, the main theoretical results concerning this process. Then, based upon these latter, we will show how one can obtain shaped THz pulses.

Let us consider an ultrashort near-infrared (NIR) laser pulse exciting, at normal incidence, a ZnTe emitter cut along the $\langle 110 \rangle$ plane, with a thickness $L_e$. Assuming a propagation in the $z$-direction, we can write its electric field in the following form:

$$E(t,x,y,z) = u(x,y,z)\mathcal{E}(t) \exp\left[i(\omega_0 t - k_0 z)\right], \tag{1}$$

where $u(x,y,z)$ and $\mathcal{E}(t)$ are, respectively, the spatial and temporal amplitudes of the laser beam, $\omega_0$ its carrier frequency (corresponding to a central wavelength $\lambda_0 = 2\pi c/\omega_0$, $c$ is the velocity of light in vaccum), and $k_0$ the mean wave vector ($k_0 = \omega_0 n(\omega_0)/c$, where $n$ is the refractive index of ZnTe). Here, we assume that this pulse is not modified during its propagation in the nonlinear crystal, so that:

$$u(x,y,z) = \mathcal{F}(x,y) \exp\left(i\frac{\omega z}{v_g}\right),$$ (2)

$v_g$ being the group velocity of the laser pulse: $v_g = c/n_g(\omega_0)$, where $n_g(\omega_0) = n(\omega_0) + \omega_0 \left(dn/d\omega\right)_{\omega_0}$ is the group index of the NIR pulse.

At each point of the medium, this pulse induces via the frequency difference between its different spectral components a second order nonlinear polarization whose spectral components $\Omega$ lie in the terahertz range. More precisely, the nonlinear polarization induced at frequency $\Omega$ is given by:

$$P^{(2)}(\Omega,x,y,z) = \varepsilon_0 \chi^{(2)}(\Omega) \int \frac{d\omega}{2\pi} E(\omega,x,y,z) E^*(\omega - \Omega,x,y,z),$$ (3)

where $\varepsilon_0$ is the vacuum permittivity, $\chi^{(2)}$ the second order nonlinear susceptibility of ZnTe, and $E(\omega,x,y,z) = \int E(t,x,y,z) \exp(i\omega t)dt$. Note that, in Eq. (3), we have assumed that the spectral dependence of $\chi^{(2)}$ in the NIR range is negligible. Taking into account Eqs. (1) and (2), and introducing the spectral amplitude of the optical field $\mathcal{E}(\omega) = A(\omega) \exp[i\phi(\omega)] = \int \mathcal{E}(t) \exp(i\omega t)dt$, one obtains:

$$P^{(2)}(\Omega,x,y,z) = \varepsilon_0 |\mathcal{F}(x,y)|^2 e^{i\frac{\Omega z}{v_g}} \chi^{(2)}(\Omega) C(\Omega),$$ (4)

where $C(\Omega)$ is the power spectrum of the rectified NIR pulse, given by:

$$C(\Omega) = \int |\mathcal{E}(t)|^2 \exp(i\Omega t)dt,$$

$$= \int \frac{d\omega}{2\pi} \mathcal{E}(\omega) \mathcal{E}^*(\omega - \Omega),$$

$$= \int \frac{d\omega}{2\pi} A(\omega) A(\omega - \Omega) e^{i[\phi(\omega) - \phi(\omega - \Omega)]}.$$ (5)

From a quantum mechanical point of view, $C(\Omega)$ is proportional to the probability amplitude that two photons of the NIR pulse with frequencies $\omega$ and $\omega - \Omega$ interact in the nonlinear crystal to generate a THz photon at frequency $\Omega$.

So, this second order polarization emits THz radiations at different frequencies $\Omega$, and, consequently, acts as a source term in the wave equation for the THz wave generated by optical rectification. This latter equation, written in the spatial and spectral Fourier domains, is given by [2, 6]

$$\left[\frac{\partial^2}{\partial z^2} + k_{\parallel}^2\right] E_{THz}(\Omega,k_x,k_y,z) = -\frac{\Omega^2}{c^2} \chi^{(2)}(\Omega) C(\Omega) \mathcal{G}(k_x,k_y) e^{i\Omega z/v_g},$$ (6)

where

$$\mathcal{G}(k_x,k_y) = \frac{1}{(2\pi)^2} \iint |\mathcal{F}(x,y)|^2 e^{-i(k_x x + k_y y)} dxdy,$$ (7)

and $k_\parallel = \sqrt{k^2(\Omega) - k_\perp^2}$, with $k_\perp = \sqrt{k_x^2 + k_y^2}$ and $k(\Omega) = n(\Omega)\Omega/c$ (wave vector of the THz wave).

At the exit of the ZnTe crystal ($z = L_e$), the amplitude of each spectral component of the THz pulse generated by optical rectification is given by [6, 7]

$$E_{THz}(\Omega, k_x, k_y, L_e) = E_{THz}(\Omega, k_\perp, L_e) = iL_e \frac{\Omega^2}{c^2} \frac{\chi^{(2)}(\Omega)}{k_\parallel(\Omega) + \Omega/v_g} e^{i[k_\parallel(\Omega) + \Omega/v_g]\frac{L_e}{2}}$$

$$\times C(\Omega)\mathcal{G}(k_\perp) \, \mathrm{sinc}\left[\frac{L_e \Delta k}{2}\right], \tag{8}$$

where $\mathcal{G}(k_\perp) = \mathcal{G}(k_x, k_y)$, and $\Delta k = k_\parallel(\Omega) - \Omega/v_g$ is the wavevector mismatch between the generated THz wave and the incident optical wave.

In Eq. (8), three terms play a key role concerning the shape of the emitted THz spectrum and the way it is related to the NIR pulse shape: the phase mismatch factor sinc $[L_e\Delta k/2]$, $C(\Omega)$ and $\mathcal{G}(k_\perp)$. Here, we will discuss only the influence of the sinc term, the discussion concerning $C(\Omega)$ and $\mathcal{G}(k_\perp)$ being postponed to sections 3 and 4 respectively. The phase mismatch factor gives its main contribution to the THz spectrum for the frequency $\Omega$ such that $\Delta k(\Omega) = 0$, the so-called phase matching condition. Assuming that the NIR pulse is almost a plane wave across the crystal, *i.e.* $k_\perp \simeq 0$, this latter condition may be written:

$$\Delta k = k(\Omega) - \frac{\Omega}{v_g} = \Omega\left(\frac{1}{v_\varphi} - \frac{1}{v_g}\right) = 0, \tag{9}$$

where $v_\varphi = c/n(\Omega)$ is the phase velocity of the THz wave. Thus, the phase matching condition sets that:

$$v_\varphi(\Omega) = v_g(\omega_0) \iff n(\Omega) = n_g(\omega_0). \tag{10}$$

Dispersion of ZnTe in the NIR range is well describe by the following Sellmeier equation [25]:

$$n^2(\lambda) = 4.27 + \frac{3.01\lambda^2}{\lambda^2 - 0.142}, \tag{11}$$

where $\lambda$ is in µm, and in the THz range by the dielectric function given in [12]:

$$\tilde{\varepsilon}(\Omega) = (n + i\kappa)^2 = \varepsilon_{el} + \frac{\varepsilon_{st}\Omega_{TO}^2}{\Omega_{TO}^2 - \Omega^2 - i2\gamma\Omega}, \tag{12}$$

with $\varepsilon_{el} = 7.44$, $\varepsilon_{st} = 2.58$, $\Omega_{TO}/2\pi = 5.32$ THz and $\gamma/2\pi = 0.022$ THz. So, a NIR pulse with a spectrum centered at $\lambda_0 \sim 800$ nm leads to a THz wave centered around 2.5 THz (see, for example, the black curve of figure 2b). Actually, the way the phase mismatch factor affects the shape of the THz spectrum depends on two factors: $\lambda_0$ and $L_e$ [1]. As a consequence, it offers little possibilities to generate shaped THz spectra with an easy tunability. However, we will see that $C(\Omega)$ and $\mathcal{G}(k_\perp)$ are more promising candidates for this purpose.

Finally, in order to compare the experimental results presented below with the theoretical predictions of Eq. (8), we have to take into account the detection of the THz electric field via electro-optic (EO) sampling in a second $\langle 110 \rangle$ ZnTe crystal, with a thickness $L_d$, using a weak probe NIR ultrashort pulse. To this end, we use the expression of the EO signal given in [11]:

$$S(\tau) \propto \int_{-\infty}^{+\infty} E_{\text{THz}}(\Omega, k_\perp, L_e) f(\Omega) e^{-i\Omega\tau} d\Omega = \int_{-\infty}^{+\infty} S(\Omega) e^{-i\Omega\tau} d\Omega, \tag{13}$$

where $\tau$ is the time delay between the THz and the NIR pulses, $S(\Omega)$ is the spectral amplitude of the EO signal, and

$$f(\Omega) = C_{\text{probe}}(\Omega) \chi^{(2)}(\Omega) \frac{e^{iL_d\Delta k} - 1}{i\Delta k}, \tag{14}$$

$C_{\text{probe}}(\Omega)$ being the power spectrum of the probe pulse.

Furthermore, in Eqs. (8) and (14), we will assume that the frequency dependence of $\chi^{(2)}$ is given by [39]:

$$\chi^{(2)}(\Omega) = 2d_{14} \left[ 1 + C \left( 1 - \frac{\Omega}{\Omega_{\text{TO}}} \right)^{-1} \right], \tag{15}$$

where $d_{14} = 90 \, \text{pm}/\text{V}$ and $C = -0.07$.

# 3. Generation of tunable THz pulses by optical pulse shaping in the spectral domain

## 3.1. Further analysis of THz generation by optical rectification I

In section 2, we have shown the influence of the phase mismatch factor on the THz spectrum. Here, we are going to study in which way one can control the shape of the THz pulse by acting on the power spectrum $C(\Omega)$. To this end, we assume that the NIR pulse is a plane wave in the crystal, so that $k_\perp = 0$ and $\mathcal{G}(k_\perp) = 1$ in Eq. (8). With these assumptions, it is clear that, apart from the sinc term, this is the power spectrum, i.e. the *temporal* pulse shape of the incident NIR pulse, since $C(\Omega) = \int |\mathcal{E}(t)|^2 \exp(i\Omega t) dt$, that controls the shape of $E_{\text{THz}}(\Omega, L_e)$. Actually, Eq. (5) clearly shows that $C(\Omega)$ strongly depends on the NIR pulse spectrum and more especially on the relative phase $\phi(\omega) - \phi(\omega - \Omega)$ between all frequency pairs within the pulse separated by $\Omega$. So, by acting on the spectral phase $\phi(\omega)$ of the NIR pulse in the appropriate way it is possible to finely control $C(\Omega)$ and, consequently, the THz spectrum shape.

In what follows, we will show some examples of tunable narrow-band THz spectra, the tunability being achieved through a sinusoidal spectral phase modulation and the spectral narrowing through an additional triangular phase.

## 3.2. Experimental setup

The experimental setup is depicted on Fig. 1: a Ti:sapphire chirped pulse amplifier (Femtopower Compact Pro) delivering laser pulses of duration $\tau_p = 35$ fs at 790 nm with a 1 kHz repetition rate is used to generate and detect THz waves. About 90% of the output of the CPA is tailored by a pulse shaper and is used for the generation of THz pulses by optical rectification in a $\langle 110 \rangle$ ZnTe crystal with a thickness $L_e$ ($L_e = 300 \, \mu\text{m}$ for the experiments of section 3.3 and 1 mm for those of sections 3.4 and 3.5).

The pulse shaper consists in a half-zero-dispersion line made of a 600 g/mm grating and a $f = 600$ mm cylindrical mirror that spatially disperses and focuses all the spectral components

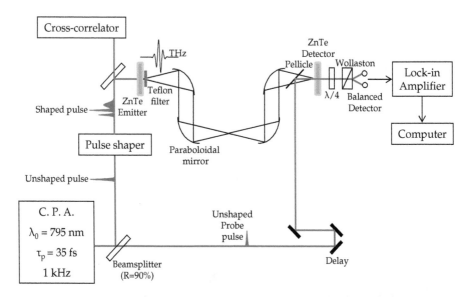

**Figure 1.** Experimental setup for the generation of tunable THz pulses by optical pulse shaping in the spectral domain.

of the pulse in the Fourier plane where a programmable single liquid crystal spatial light modulator (LC SLM) with 640 pixels, manufactured by Jenoptik AG, acts as a spectral filter [22, 23, 33, 42]. A plane mirror positioned just after the SLM folds the line to perform an inverse Fourier transform and reassemble all the spectral components of the shaped pulse. The SLM is such that it is only possible to change the spectral phase of the incoming pulse. The intensity profile of the shaped laser pulse is measured by cross-correlation, in a type I BBO crystal, of this pulse with an unshaped reference pulse.

The beam exiting from the pulse shaper then excites the ZnTe emitter at normal incidence. After removal of the residual NIR pulse by a Teflon filter, the generated THz wave is imaged into a second $\langle 110 \rangle$ ZnTe crystal with a thickness $L_d$ by means of four off-axis paraboloidal mirrors ($L_d = 300$ μm for the experiments of section 3.3 and 500 μm for those of sections 3.4 and 3.5). There, it is measured via electro-optic effect using a weak unshaped NIR pulse coming directly from the exit of the Ti:sapphire amplifier [11]. By varying the delay $\tau$ between the THz pulse and the ultrashort NIR pulse, one gets an EO signal, $S(\tau)$, corresponding to a cross-correlation between the laser pulse and the THz wave integrated over the EO crystal length. This signal is then recorded by ellipsometry (the ellipsometer consists in a $\lambda/4$ wave plate, a Wollaston polarizer, and two balanced detectors).

### 3.3. Sinusoidal phase modulation: generation of THz pulse trains

It is well known that selectivity in the THz spectral range can be achieved by means of terahertz-rate sequences of femtosecond pulses [1, 21, 41]. Experimentally, such a pulse train with a repetition rate $F_{rep}$ is generated by applying a periodic phase modulation with a period $F_{rep}$, the most used being $\phi(\omega) = a \sin(\omega/F_{rep})$. In the absence of any phase modulation, *i.e.*

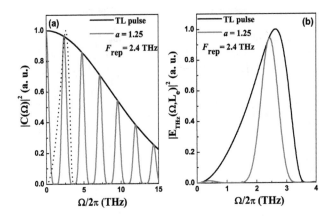

**Figure 2.** (a) Power spectrum of an optical pulse with $\lambda_0 = 790$ nm, and a pulse duration $\tau_p = 35$ fs: transform-limited (TL) pulse (black curve), shaped pulse with a sinusoidal phase modulation $\phi(\omega) = a \sin(\omega/F_{rep})$ (red curve). The dotted curve superimposed to $|C(\Omega)|^2$ corresponds to the THz spectrum generated by a TL pulse. (b) Terahertz spectra generated by optical rectification in a $\langle 110 \rangle$ ZnTe of a: TL pulse (black curve), sinusoidally phase modulated pulse (red curve).

for a transform-limited pulse corresponding to $\phi(\omega) = 0$, the power spectrum is a bell-shaped curve centered on $\Omega = 0$, with a half-width at maximum equal to the pulse bandwidth (black curve, Fig. 2a). However, if a sinusoidal phase modulation is applied, then $C(\Omega)$ exhibits spectral interference fringes with maxima centered at the frequencies $\Omega_n = 2\pi n F_{rep}$ (where $n$ is an integer) (red curve, Fig. 2a, where $a = 1.25$ and $F_{rep} = 2.4$ THz). Now, if one takes into account the spectral filtering of the powerspectrum induced by the phase matching condition, corresponding to a multiplication of $|C(\Omega)|^2$ by the dotted curve of Fig. 2a, it results in a narrower THz spectrum centered at the frequencies $\Omega_n$ within the THz band authorized by the phase matching condition (Fig. 2b). This spectral narrowing can also be understood by a reasoning in the time domain. Actually, each pulse of the NIR pulse train, two consecutive pulses being separated by $T_{rep} = 1/F_{rep}$, leads to its own THz waveform. All these waveforms add coherently to generate a THz train with a repetition rate $F_{rep}$ whose spectrum precisely consists in peaks centered at frequencies $\Omega_n = 2\pi n/T_{rep}$, the bandwidth of these peaks decreasing as the number of pulses in the THz train increases.

Fig. 3a displays the EO signals, $S(\tau)$, associated to THz trains generated with our setup for different values of $F_{rep}$ and $a = 1.25$, whereas the corresponding spectra, defined as $|S(\Omega)|^2$, are represented in Fig. 3c. In agreement with the former analysis, we observe a THz pulse train with a spectrum that can be tuned within the spectral range obtained for an unshaped pulse by adjusting the repetition rate of the NIR pulse train. There is a very good agreement between our results and the spectra calculated with the model described in section 2 (Fig. 3c and d). These two last figures clearly show that we can tune the carrier frequency of THz pulse simply by changing the repetition rate of the NIR pulse train.

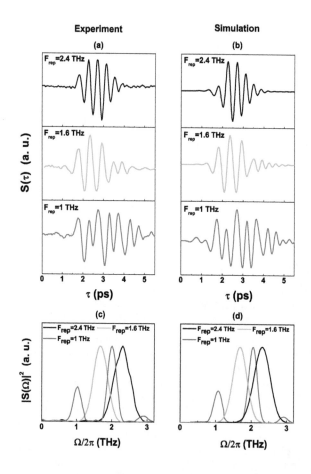

**Figure 3.** THz pulse train with an adjustable repetition rate $F_{rep}$, the paramater $a$ being equal to 1.25. Experiment: (a) waveforms and (c) corresponding spectra. Simulation: (b) waveforms and (d) corresponding spectra (taken from [37]).

However, for $F_{rep} = 1$ THz, one can note that the spectrum also displays second and third-harmonic peaks centered at 2 and 3 THz respectively. They are inherent to the use of a sinusoidal phase modulation: for $F_{rep} = 1$ THz, the phase matching condition is less selective with respect to the peaks of the power spectrum, leading to these harmonics. From a "temporal" point of view, these harmonics arise because, for this repetition rate, the THz waveform generated by each NIR pulse of the train has a duration shorter or comparable to $T_{rep}$. Thus, when all these partial waveforms add coherently to generate the total THz electric field $E_{THz}(t)$, they leads to an overall shape that is more complex and damps more slowly than the sinusoidal waveform obtained for $F_{rep} = 1.6$ or 2.4 THz. Thanks to the higher

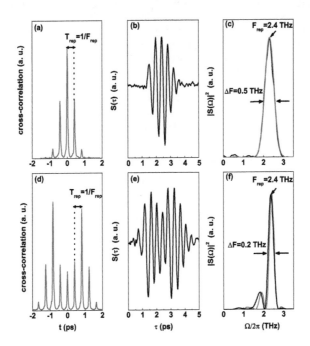

**Figure 4.** Experimental illustration of the spectral narrowing by a sinusoidal phase modulation $\phi(\omega) = a\sin(\omega/F_{rep})$: (a) cross-correlation of the optical pulse train generated with $a = 1.25$ and $F_{rep} = 2.4$ THz. (b) Associated THz train and (c) corresponding spectrum: experiment in black and simulation in red. (d) cross-correlation of the NIR pulse train generated with $a = 2.65$ and $F_{rep} = 2.4$ THz. (e) Associated THz train and (f) corresponding spectrum (taken from [37]).

number of oscillations, that is to the longer THz pulse duration, the width of the harmonics for $F_{rep} = 1$ THz is also reduced ($\sim 280$ GHz) compared to the spectra for $F_{rep} = 1.6$ or 2.4 THz (bandwidth $\sim 500$ GHz).

To see if it is possible to further decrease the bandwidth of the THz spectrum when a sinusoidal phase modulation is applied to the rectified NIR pulse, let us consider a transform-limited laser pulse with a duration $\tau_p$, and a gaussian amplitude

$$A(\omega) = E_0 T_p \sqrt{\pi} \exp\left(-\frac{\omega^2 T_p^2}{4}\right), T_p = \tau_p / \sqrt{2\ln 2}. \tag{16}$$

When a sinusoidal phase $\phi(\omega)$ is applied to this pulse, it results in a pulse train with a temporal dependence of the electric field given by [29]:

$$\mathcal{E}(t) \propto \sum_{m=-\infty}^{+\infty} J_m(a) \exp\left(\frac{im\pi}{2}\right) \exp\left[-\left(\frac{t - mT_{rep}}{T_p}\right)^2\right], \tag{17}$$

where the $J_m$ are Bessel functions of the first kind. This equation shows that, *whatever* the phase modulation period $F_{rep}$, the amplitude of the $m^{th}$ pulse of the train is given by $|J_m(a)|$. So, by playing on the modulation strength $a$ one can control, to some extend, the weight of the different pulses of the train. More precisely, the evolution of $|J_m(a)|$ is the following [29]: except for $|J_0|$, it increases when $a$ is varied from 1.25 to 2.65, leading to an enhancement of the number of pulses in the train. Hence, increasing $a$ will lengthen the THz pulse duration and lead to a narrower spectrum, this regardless of the repetition rate $F_{rep}$. Fig. 4f displays the THz spectrum obtained by applying a sinusoidal phase modulation with $a = 2.65$ and $F_{rep} = 2.4$ THz. With this modulation strength, there are nine pulses in the NIR pulse sequence (Fig. 4d) instead of five for $a = 1.25$ (Fig. 4a), giving a spectrum with a bandwidth of 0.2 THz. Here again, experiment and simulation are in good agreement (Fig. 4f).

Nevertheless, one also observes some ripples on the spectrum of Fig. 4f. They can be attributed to the difference in amplitude between the different pulses of the NIR pulse train, and especially to the two main peaks centered around $\pm 800$ fs in Fig. 4d, resulting from the fact that, for $a = 2.65$, $|J_{\pm 2}(a)| > |J_{\pm n}(a)|$ ($|n| \neq 2$). Unfortunately, their amplitude cannot be modified easily without affecting the amplitude of the other peaks. Indeed, as shown in reference [29], it is impossible to produce pulse trains with more than three pulses of equal amplitude with a simple sinusoidal phase. On the other hand, one can solve this problem by adding higher harmonic orders to the applied phase modulation, *i.e.* by means of the following spectral phase: $\phi(\omega) = \sum_n F_n \sin\left(n\omega/F_{rep} + \phi_n\right)$. However, this phase is harder to implement since it requires the use of an optimization algorithm to find the best value for the weight $F_n$ of each harmonic. In the same spirit, Ahn et al. [1] have used a Gerchberg-Saxton algorithm to find the best spectral phase in order to generate many nearly equally spaced identical optical pulses. But, here again, this method relies on the ability of the algorithm to converge towards the expected solution. Finally, we will show in section 3.5 that it is possible to shrink the THz pulse bandwidth simply by adding a soundly chosen triangular phase modulation to the sinusoidal one. Before that, we will consider the effect of a triangular phase modulation on the THz generation.

### 3.4. Triangular phase modulation: generation of phase-locked THz pulse pairs

Let us consider a triangular phase modulation defined by

$$\phi(\omega) = -\Delta\tau|\omega - \delta\omega|, \tag{18}$$

where $\Delta\tau$ is the slope of the spectral phase, and $\delta\omega$ is the detuning with respect to the center of the spectral amplitude $\mathcal{E}(\omega)$. Such a phase, when it is applied to a transform-limited pulse with a duration $\tau_p$ and a gaussian shape

$$\mathcal{E}(t) = E_0 \exp\left[-\left(\frac{t}{T_p}\right)^2\right], T_p = \tau_p/\sqrt{2\ln 2}, \tag{19}$$

leads to a pair of pulses whose temporal amplitude is given by [29]:

$$\mathcal{E}(t) = \mathcal{E}_+(t + \Delta\tau) + \mathcal{E}_-(t - \Delta\tau), \tag{20}$$

with

$$\mathcal{E}_\pm(t) = \frac{E_0}{2} e^{-(t/T_p)^2} \left[1 - \mathrm{erf}\left(i\frac{t}{T_p} \pm \frac{T_p\delta\omega}{2}\right)\right], \tag{21}$$

with the error function, erf, defined by

$$\text{erf}(x) = \frac{2}{\sqrt{\pi}} \int_0^x \exp(-y^2)dy. \tag{22}$$

So it leads to two pulses separated by a time delay $t_p = 2\Delta\tau$, with a ratio of their amplitudes $(\mathcal{E}_+(0)/\mathcal{E}_-(0))$ controlled by the detuning $\delta\omega$, this ratio being equal to 1 for $\delta\omega = 0$. This kind of sequence of ultrashort phase-locked pulses has been used extensively to perform temporal coherent control experiments with optical pulses [3]. It has also been used successfully in the terahertz range to control various processes [15, 16, 44] (see also section 3.5 for a brief presentation). In these latter experiments, the THz pulse pair was generated from a NIR pulse pair created by a Michelson or a Mach-Zender interferometer. Here, we will demonstrate that it is possible to generate a THz pulse pair by applying a triangular spectral phase to the rectified NIR pulse. Since temporal coherent control experiments require a balanced pair of pulses to achieve the best efficiency, we will only consider, in what follows, the case where $\delta\omega = 0$, so that $\mathcal{E}_+(t) = \mathcal{E}_-(t)$.

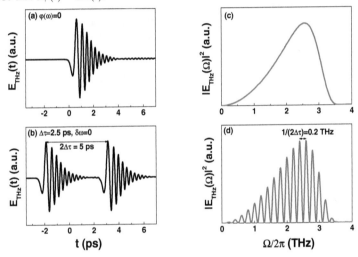

**Figure 5.** THz waveforms generated by optical rectification in a 300 µm-thick ZnTe crystal of: (a) a transform-limited NIR pulse, (b) a NIR pulse having a triangular phase modulation with $\Delta\tau = 2.5$ ps and $\delta\omega = 0$. (c) and (d) respectively show the spectra of (a) and (b). In (d), the fringe spacing is $1/t_p = 0.2$ THz.

By optical rectification of a pair of balanced NIR pulses in a ZnTe crystal, one obtains a THz pulse pair with the following electric field:

$$E_{THz}(t) = E_{THz_+}(t + \Delta\tau) + E_{THz_+}(t - \Delta\tau), \tag{23}$$

where $E_{THz_+}(t)$ is the electric field of the THz pulse generated by optical rectification of a NIR pulse whose temporal amplitude is $\mathcal{E}_+(t)$. Its amplitude spectrum is given by

$$E_{THz}(\Omega) = E_{THz_+}(\Omega)e^{-i\Omega\Delta\tau} + E_{THz_+}(\Omega)e^{i\Omega\Delta\tau}$$
$$= E_{THz_+}(\Omega)e^{-i\Omega\Delta\tau}\left(1 + e^{i\Omega t_p}\right), \tag{24}$$

so that the expected spectrum of the EO signal is

$$|S(\Omega)|^2 \propto |E_{\text{THz}}(\Omega)|^2 = 2|E_{\text{THz}_+}(\Omega)|^2 \left[1 + \cos\left(\Omega t_p\right)\right]. \tag{25}$$

It has the same enveloppe as the spectrum $|E_{\text{THz}_+}(\Omega)|^2$ of a single THz pulse (shown on Fig. 5c), but with modulations due to the spectral interference of the two THz pulses, the fringes spacing being $1/t_p$ (see Fig. 5d).

Fig. 6 shows the THz pulse pair generated experimentally by applying a triangular phase modulation with $\Delta\tau = 2.5$ ps and $\delta\omega = 0$ to the NIR pulse (the same parameters as in the simulation of Fig. 5). As mentioned previously, for $\delta\omega = 0$, both NIR pulses are balanced (Fig. 6d), and give rise to a pair of THz pulses with the same carrier frequency and separated by $t_p = 5$ ps (Fig. 6e). The corresponding EO signal spectrum (Fig. 6f) has the same enveloppe as the EO signal spectrum of a single THz pulse (Fig. 6c), but exhibits oscillations with a period $1/t_p = 0.2$ THz.

However, in addition to the NIR pulse pair, the cross-correlation trace of Fig. 6d also has an unwanted TL pulse centered at $t = 0$. This distortion is due to the combination of two limitations of the LC SLMs [35]: first, since LC SLMs have a phase operating range slightly above $2\pi$, the applied phase modulation is "wrapped" to be within the $[0, 2\pi]$ interval

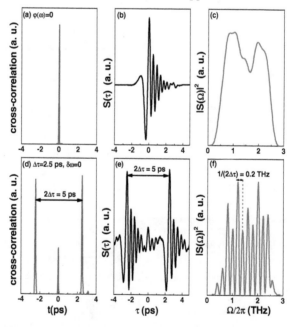

**Figure 6.** Sequence of THz phase-locked pulses experimentally generated by applying a triangular phase modulation with $\Delta\tau = 2.5$ ps and $\delta\omega = 0$ to the rectified NIR pulse: (a) cross-correlation of the initial TL NIR pulse, (b) associated THz waveform and (c) corresponding spectrum. (d) cross-correlation of the NIR pulse pair, (e) associated THz waveform and (f) corresponding spectrum (taken from [37]).

[33]. Second, the pixels, and consequently the gaps, of the LC SLM are not sharply defined, leading to some smoothing of the pixelated phase. Combined together these limitations lead to the apparition of "modulator replica" pulses in addition to the desired pulse shape. In our experiment, the phase varies from $-223\pi$ to $0$ over the first half of the spectrum and, symetrically, from $0$ to $-223\pi$ over the second half. This leads to numerous phase-wraps and, in combination with the smoothed spatial response of the pixels, gives the pulse centered at $t = 0$ in Fig. 6d. This unwanted effect can be avoided by the use of a diffraction-based pulse shaper [36]. Nevertheless, as we will see in the next section, it does not affect significantly the THz spectrum shape.

### 3.5. Sinusoidal + triangular phase: tunable narrow band THz pulses.

We now consider the action of both phases introduced previously, i.e.

$$\phi(\omega) = a\sin(\omega/F_{rep}) - \Delta\tau|\omega - \delta\omega|. \tag{26}$$

According to sections 3.3 and 3.4, we expect a sequence of pulse trains with a repetition rate $F_{rep}$ separated by a delay $t_p = 2\Delta\tau$. This is confirmed by Fig. 7a obtained for $a = 1.25$, $F_{rep} = 2$ THz, $\Delta\tau = 2.5$ ps and $\delta\omega = 0$. As in section 3.4, the applied triangular phase modulation leads to numerous phase-wraps and, in combination with the limitations of the LC SLM, gives an additional TL pulse centered at $t = 0$. This undesired pulse and the pulse train sequence generate their own THz waveforms that add coherently (Fig. 7b). The THz spectrum is displayed on Fig. 7c. Due to the triangular phase modulation, it still exhibits interference fringes with a spacing equal to $1/t_p$, but, thanks to the sinusoidal phase, it is narrower than the one of Fig. 6f and is centered in $F_{rep}$. Note that it is not affected significantly by the THz pulse generated by the TL pulse due to the limitations of the SLM as can be seen from the simulation of Fig. 7f which does not include these latter.

These THz pulse trains sequences are of great importance for temporal coherent control and the generation of tunable narrow-band THz pulses. The goal of temporal coherent control is to excite selectively some levels of a quantum system by means of a pair of ultrashort phase-locked pulses in order to steer the evolution of this system towards a desirable state through its interaction with light (for example to control the outcome of a photo-induced chemical reaction). Let us consider, for example, a quantum system with a ground state $|g\rangle$ and $N$ excited states $|e_k\rangle$ (with energies $\hbar\Omega_k$), which is excited by a sequence of two identical ultrashort pulses separated by a delay $t_p$. We suppose that the bandwidth of the pulses is broad enough to excite the $N$ states $|e_k\rangle$. In the weak field regime, the state vector $|\Psi(t)\rangle$ of the system after interaction with this sequence of pulses is given by [3]:

$$|\Psi(t)\rangle = |g\rangle + \sum_{k=1}^{N} b_k e^{-i\Omega_k t}|e_k\rangle, \tag{27}$$

with

$$b_k = \frac{1}{i\hbar}\mu_{kg}E_{tot}(\Omega_k), \tag{28}$$

where $\mu_{kg}$ is the transition dipole moment from state $|g\rangle$ to state $|e_k\rangle$, and $E_{tot}(\Omega_k)$ is the spectral amplitude of the total electric field taken at frequency $\Omega_k$, given by

$$E_{tot}(\Omega_k) = E(\Omega_k)\left(1 + e^{i\Omega_k t_p}\right), \tag{29}$$

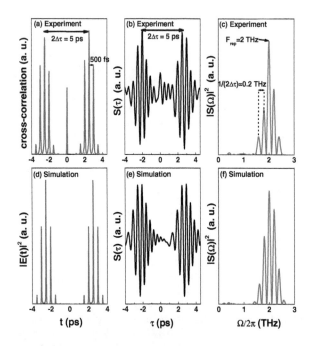

**Figure 7.** Sequence of THz pulse trains generated by applying simultaneously a triangular and a sinusoidal phase with $a = 1.25$, $F_{rep} = 2$ THz, $\Delta\tau = 2.5$ ps and $\delta\omega = 0$. (a) cross-correlation of the NIR pulse, (b) associated THz waveform and (c) corresponding spectrum. Panels (d), (e) and (f) correspond respectively to the simulations of panels (a), (b) and (c) (taken from [37]).

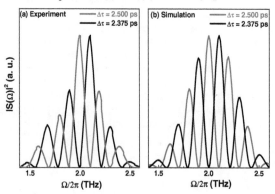

**Figure 8.** "In-phase" and "out-of-phase" sequences of pulse trains at $\Omega_k/2\pi = 2$ THz: (a) Experimental THz spectra obtained for $a = 1.25$, $F_{rep} = 2$ THz, $\delta\omega = 0$, $\Delta\tau = 2.5$ ps (red curve, corresponding to $\Delta\varphi = 2\pi p$) and $\Delta\tau = 2.375$ ps (black curve, corresponding to $\Delta\varphi = 2\pi(p + 1/2)$). (b) Corresponding simulations (taken from [37]).

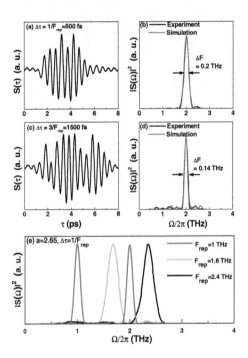

**Figure 9.** Spectral smoothing ((a) and (b)) and spectral narrowing ((c) and (d)) with a sequence of THz pulse trains. In both cases $a = 2.65$, $F_{rep} = 2$ THz and $\delta\omega = 0$. (e) Experimental illustration of the tunability of the spectral smoothing shown in (a): for all the curves $a = 2.65$, $\Delta\tau = 1/F_{rep}$ and $\delta\omega = 0$ (taken from [37]).

$E(\Omega)$ being the amplitude spectrum of a single pulse. Thus, the probability to excite the state $|e_k\rangle$ is

$$|\langle e_k|\Psi(t)\rangle|^2 = |b_k|^2 \propto |E(\Omega_k)|^2 \left[1 + \cos\left(\Omega_k t_p\right)\right]. \tag{30}$$

This probability strongly depends on the relative phase $\Delta\varphi(\Omega_k) = \Omega_k t_p$ between the two pulses of the sequence: the excitation is "constructive" for $\Delta\varphi = 2\pi p$ ($p$ is an integer) and "destructive" for $\Delta\varphi = 2\pi(p + 1/2)$. The fine tuning of $\Delta\varphi$ is achieved via the tuning of the delay $t_p$, and strongly depends on its (interferometric) stability. In the THz range, our setup makes it possible to generate easily these "in-phase" and "out-of-phase" pulse pairs, simply by changing the slope $\Delta\tau$ of the triangular phase. For instance, Fig. 8 shows two THz pulse sequences leading respectively to a constructive excitation (red curve) and a destructive excitation (black curve) at $\Omega_k/2\pi = 2$ THz.

One can also take advantage of the combined triangular and sinusoidal phases to generate less structured THz spectra than the one of Fig. 4f, simply by choosing a delay between the THz pulses shorter than their duration. Then the THz waveforms are superposed and, for a suitably choosen value of $\Delta\tau$, give rise to a single THz train with pulses having almost the same amplitude, resulting in a spectrum with a well defined peak, centered at the pulse train repetition rate $F_{rep}$. This remarkable value of $\Delta\tau$ is $1/F_{rep}$, $\forall F_{rep}$. Figures 9a and 9b display an example of such a "smooth" narrow-band THz spectrum for $F_{rep} = 2$ THz, whereas Fig. 9e shows the tunability of this method. On this figure, one can see that, contrary to Fig. 3c, the peaks at 1 and 2 THz for $F_{rep} = 1$ THz have the same amplitude. It results from the fact that this spectrum was obtained with a thicker crystal than the one used in section 6, leading to an unshaped THz spectrum with almost the same energy around 1 and 2 THz (see Fig. 6c) unlike the unshaped spectrum corresponding to Fig. 3. Now, for a given value of $F_{rep}$, if one tunes $\Delta\tau$ around $1/F_{rep}$, it is then possible to generate relatively clean THz spectra with a narrower bandwidth. An example is given in figures 9c and 9d: for $\Delta\tau = 3/F_{rep} = 1.5$ ps, the THz waveform exhibits more oscillations than in Fig. 9a, leading to a spectrum with a bandwidth of 140 GHz. However, for repetition rates like $F_{rep} = 1$ THz, one should note that although this method makes it possible to narrow and smooth the peak at 1 THz and its harmonics, it cannot suppress these latter since they are inherent to the use of a sinusoidal phase modulation.

## 3.6. Limitations of this method

In this section, we will address some points concerning the limitations of our approach and the possibility to generate even narrower THz spectra.

First, we will consider the limitations imposed by the pulse shaper. We have already met such limitations in sections 3.4 and 3.5, leading to the apparition of "modulator replica" pulses in addition to the desired pulse shape. Now, we will see that the pixelated nature of the LC SLM imposes upper limits to $T_{rep}$ and $\Delta\tau$. Indeed, to be correctly sampled by the modulator's pixels, and thus avoid temporal aliasing, the phase modulation must satisfy the Nyquist's criterion: the increase of phase between two adjacent pixels of the SLM must not exceed $\pi$ [22, 33, 42]. For the sinusoidal and triangular phase modulations considered separately, this criterion, applied to our setup, leads respectively to $T_{rep} \leq 4$ ps (i.e. $F_{rep} \geq 250$ GHz) and $\Delta\tau \leq 4$ ps (i.e. $t_p \leq 8$ ps). If these conditions are not satisfied, then sampling replica pulses alter the desired output waveform [22, 35]. So, the characteristics of the pulse shaper limit the ability to generate smooth and tunable THz spectra at low frequencies, that is below 500 GHz. Nevertheless, these frequencies are not of great concern here since the THz spectrum generated by optical rectification in this spectral range is almost zero.

Actually, the main limiting factor of our approach, especially if one intends to generate THz spectra with a bandwidth below 100 GHz, is the use of a simple sinusoidal phase modulation. As already mentioned in section 3.3, one can circumvent this limitation by adding harmonics to the applied phase, the number of harmonics increasing inversely with the required THz bandwidth. Indeed, an enhancement of the number of harmonics makes it possible to increase the number of pulses with the same amplitude in the NIR pulse train and, consequently, to reduce the bandwidth of the generated THz spectrum. However, one must not forget that the determination of the coefficients $F_n$ and $\phi_n$ requires the use of optimization methods like the one of reference [1]. Moreover, it is important to have in mind that in LC SLM pulse shapers, the tailored output pulse reflects accurately the response of the SLM over a restricted temporal window. The duration $T$ of this window is governed either by the finite spectral resolution of the zero-dispersion line or by the pixelation of the SLM (i.e. the Nyquist's criterion)[23, 35, 42]. In our setup, $T$ is limited by the pixelation of the SLM, and $T = 8$ ps, so the whole NIR pulse train must not exceed this duration, imposing an upper limit to the number $m$ of pulses in the train. Indeed, a train of $m$ pulses, with a repetition rate $F_{rep}$, has a duration $\Delta t$ given by $(m-1)/F_{rep}$. If $\Delta t \leq T$, then

$$m \leq m_{max} = 1 + T \times F_{rep}. \tag{31}$$

For $F_{rep} = 0.5$, 1 and 2 THz, $m_{max}$ is respectively equal to 5, 9 and 17 pulses. If we consider a pulse train with a repetition rate $F_{rep}$ and a number of identical (same amplitude and same duration) pulses equaling the maximum number of pulses $m_{max}$ authorized by the temporal window $T = 8\,ps$ for this repetition rate, a numerical calculation of the corresponding power spectrum $C(\Omega)$ shows that the spectral interference fringes induced by this pulse train (centered at $\Omega_n = 2\pi n F_{rep}$) have a bandwidth $\sim 100$ GHz, whatever the value of $F_{rep}$. It means that, with our setup, we can't generate narrow-band THz pulse with a bandwidth below 100 GHz. To go below this limit, it is necessary to work with transform-limited NIR pulses having a longer duration, the number of pixels of the SLM being unchanged. Indeed, the temporal window $T$ increases as the bandwidth of the input NIR pulse decreases [35]. For example, in reference [22], the input pulse duration is 130 fs and the LC SLM pulse shaper has a time window $T = 28$ ps (the number of pixels is 640 like in our pulse shaper). With these values, it is possible to enhance $m_{max}$, and, consequently, to generate narrow-band THz spectra with bandwidths down to $\sim 30$ GHz for all repetition rates as shown by numerical simulations. However, the tuning range is reduced, since working with longer NIR pulses decreases the width of the power spectrum, and consequently, the support of the THz spectrum.

## 4. Generation of tunable THz pulses by optical pulse shaping in the spatial domain

### 4.1. Further analysis of THz generation by optical rectification II

Our problem is to propose a way to spectrally tune the THz field generated by optical rectification in a zinc blende crystal by adjusting the transverse beam profile of femtosecond pump laser pulse. To solve this problem we will exploit the intermediate field properties of the emitted THz pulse. In order to describe the THz radiation properties in the intermediate field, we will use the Fresnel diffraction theory. To further simplify our analysis, we will consider that the beam profile has a radial symmetry and is only a function of $r$. Assuming a propagation in the $z$-direction, we will slightly modify Eq. (1) and write the electric field in the following form:

$$E(t,x,y,z) = E(t,r,z) = u(r,z)\mathcal{E}(t)\exp\left[i(\omega_0 t - k_0 z)\right], \tag{32}$$

where $r = \sqrt{x^2 + y^2}$ and $u(r,z)$ and $\mathcal{E}(t)$ are, respectively, the spatial and temporal amplitudes of the laser beam. Here again, we assume that this pulse is not modified during its propagation in the nonlinear crystal. After its generation in a very thin nonlinear crystal located at $z = z_0$, the THz electric field $E(\Omega, r, z)$ at the transverse plane $z$, is the convolution of the THz field $E(\Omega, r, z_0)$ leaving the crystal with the Fresnel propagator

$$\frac{\exp\left(ik(\Omega)(z-z_0)\left[1 + \frac{r^2}{2(z-z_0)^2}\right]\right)}{i\lambda(z-z_0)} \tag{33}$$

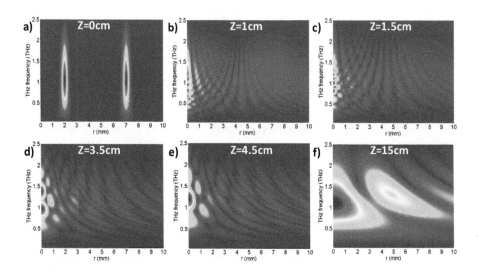

**Figure 10.** Numerical spatiotemporal behavior of a THz pulse emitted by a ZnTe crystal at different position along the propagation direction. The near I.R. femtosecond pump pattern is represented by two concentric circles that have a radius of 2 mm and 7 mm and a thickness of 200 μm (taken from [7])

where $\Omega$ and $\lambda$ are respectively the THz frequency and the associated wavelength. In the spatial frequency domain it reads:

$$E(\Omega, k_\perp, z) = E(\Omega, k_\perp, z_0) \cdot \mathbb{F}^2 \left( \frac{\exp\left(ik(\Omega)(z - z_0)\left[1 + \frac{r^2}{2(z-z_0)^2}\right]\right)}{i\lambda(z - z_0)} \right). \qquad (34)$$

The expression of $E(\Omega, k_\perp, z_0)$ is given by Eq. (8). If the pump beam has a radial symmetry, the function $G(k_\perp)$ is given by:

$$G(k_\perp) = 2\pi \int_0^\infty F(r) J_0(k_\perp r) r dr. \qquad (35)$$

It represents the Hankel transform of the pump laser fluence $F(r)$, $J_0(k_\perp r)$ being the $0^{th}$ order Bessel function. The latter equation makes it possible to compute the spatial and spectral profiles of a THz pulse propagating in a linear medium after its generation in a nonlinear crystal.

To appreciate the possibilities offered by our approach of the near field, the spatiotemporal evolution of the THz beam generated by a spatially patterned femtosecond laser pulse is shown in Fig. 10. The spatial profile of the pump beam used to generate the THz pulse in a 1 mm-thick ZnTe crystal consists in two concentric circles having a radius of 2 mm and 7 mm. At the exit of the crystal, the THz and pump beam pattern matches. However as shown in Fig. 10, along short distances, as the THz pulse propagates in free space most of its energy is confined around the z-axis ($r = 0$). In fact, in the $r$-THz frequency plane, one can observe along $r = 0$ the modulation of the THz spectrum. To account for this behavior, we have proposed the following phenomenological interpretation. At the exit of the crystal, each NIR circle simultaneously generates a THz pulse. The THz pulses are emitted as Cherenkov-radiation-like conical beams. This is partly due to the thickness of each circle which is about the THz central wavelength (or much smaller for lower THz frequencies) [4, 32]. Under these conditions, emitted THz pulses can cross themselves in the intermediate field, generating a THz pulse train in a given point along the longitudinal axis. As shown in Fig. 10, under favorable geometries, spectral interference are expected along the propagation axis.

## 4.2. Comparison between experimental results and theory.

In order to check the validity of our approach we have used the setup shown in Fig. 11. The pump pulse (about 100 μJ, 800 nm, 50 fs at 1 kHz) generates THz pulses in the first ZnTe crystal. The THz spectrum generated in the latter crystal is then detected in a second ZnTe crystal. The measurements are performed along the axis for different distance $d$ between the emitting and detecting ZnTe crystals. Both crystals were 1 mm-thick. The NIR pump and probe beams are separated by mean of a thin undoped silicon wafer. The latter is almost transparent to the THz pulse. As shown in Fig. 11 to pattern the pump beam profile we used a liquid crystal Spatial Light Modulator (SLM). The pump beam reflected by the SLM that carries the spatial phase added by the SLM is focussed by means of a 1 m focal length lens. At the focal point of the lens, where one records the spatial Fourier transform of the pump beam, we place the first ZnTe crystal. This device makes it possible to impinge patterned pump beam profiles in the ZnTe crystal that generates the THz pulse.

In the first set of experiments, THz pulse spectra are recorded along the propagation axis, under a two concentric NIR circles cross-section geometry, at three different distances from the generating crystal ($z$ = 1.5 cm, 2.5 cm and 4.5 cm). The two NIR circles have radii of 2 mm and 4 mm. The result is shown in Fig. 12. Note that the spectrum shape changes completely as the position of the detecting crystal changes, in accordance with the theoretical considerations above. This behavior can be explained by the theoretical model. In fact, the optical path difference between the two THz pulses, simultaneously emitted by the two NIR circles, decreases as the detection distance increases along the $z$-axis. As a consequence the number of interference fringes in the spectrum decreases and the spectral peak moves at the

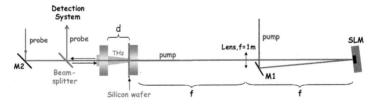

**Figure 11.** Experimental setup. Two 1 mm-thick ZnTe crystals are used to generate and detect the THz pulse. The pump and probe pulses are delivered by a regenerative amplifier (Coherent Legend) at 800 nm, with 50 fs duration, working at a 1 kHz repetition rate. The pump beam is spatial patterned by a SLM working in reflection. The THz pulse is measured along the propagation direction for different distances $d$ between the generating and detecting crystal. The pump beam is focussed in the generating crystal by means of a 1 m focal lens. The latter also realizes the spatial Fourier transform (taken from [7]).

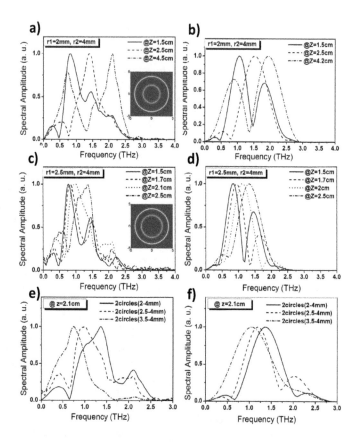

**Figure 12.** Experimental results (figures 3a, 3c, 3e) and corresponding numerical simulations (figures 3b, 3d, 3f) for THz spectra obtained by optical rectification in ZnTe under two concentric circles configurations of the pump laser beam profile. The insets show input beam rings. Each time, spectral amplitudes are normalized (taken from [7]).

same time toward the higher frequencies. A similar result is obtained for two concentric NIR circles with radii of 2.5 mm and 4 mm, and it is shown in Fig. 12c. Very good agreement is found between experiment (Fig. 12a and c) and numerical simulations (Fig. 12b and d). In the second set of experiments the detecting-ZnTe is placed at a fixed distance of 2.1 cm from the source-ZnTe. We still use a two concentric circles configuration for the pump geometry fixing the radius of the external circle at 4 mm but changing the radius of the internal circle from 2 mm to 3.5 mm. Spectra recorded for three different diameter values are shown in Fig. 12e. Note that the spectral peak moves toward the lower frequencies as the diameter of the internal circle increases. Here again, the numerical simulations (Fig. 12f) are in very good agreement with the experimental results (Fig. 12e).

Another important characteristic of this source is that most of the emitted THz radiation is spatially confined around the $z$-axis, during propagation, as shown by the numerical

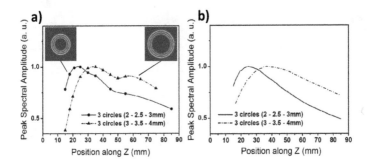

**Figure 13.** THz peak spectral amplitude against the position of the detecting ZnTe crystal along $z$. THz radiation shows interesting "self-focusing" properties. The "focal" position changes with the radius values of the circles. Fig. 4a Experiment (the insets show representative ring configuration, lines are guides for eyes); Fig. 4b simulations (taken from [7]).

simulation of Fig. 10. This behavior can be explained by the following simple argumentation. The $z$-axis is the zone of space which minimizes the optical path difference between the pulses coming from each NIR circle. Therefore constructive spatial interference between the spectral components of the THz pulses generated by each NIR circle occur mainly along the propagation axis. Furthermore, for a given cross-section geometry, the peak spectral amplitude exhibits a maximum value at a given point $z$ (the focal point). The device acts as a tunable chromatic lens. This behavior is shown in the experimental curves of Fig. 13a, obtained using a three concentric circles geometry for the pump fluence. We study two configurations. The first one is represented by three circles with radii of 2 mm, 2.5 mm and 3 mm respectively (black solid circles); in the second configuration the three circles have radii of 3 mm, 3.5 mm and 4 mm respectively (black solid triangles). In the first case the peak spectral amplitude exhibits a maximum around $z = 20$ mm, in the second case the maximum is found around $z = 35$ mm. Corresponding numerical simulations, in Fig. 13b, are in very good agreement with the experimental results.

## 4.3. Generation of spectrally shaped THz pulses: theoretical approach.

In the previous sections, we have shown that the use of spatially patterned pump pulses strongly impacts the spectral shape of the near field generated THz pulses. In fact numerical simulations based on Eq. (34) indicate that very interesting spectrally shaped THz pulses can be achieved under optimized experimental conditions. In Fig. 14 are shown results predicted by the model under two "opposite" geometrical conditions: geometries with few concentric circles widely separated between them (Fig. 14a) and geometries with many close concentric circles (Fig. 14b). In the first case (two circles) the spectrum exhibits a large number of modes, which can be translated along the frequency axis by changing the detection position along $z$ and/or the circle diameters. In the second case (five circles) a very narrow-band THz spectrum can be obtained, and its peak can be translated along the frequency axis by changing the detection position along $z$ and/or the circle diameters. From the experimental point of view, the ZnTe crystal size is a major issue for fine tuning the spectral and spatial shapes of the THz pulses. In our experiments, the maximum diameter of the circle diameter was limited by the crystal diameter (10 mm). For instance, the result in Fig. 14a and b can only be achieved

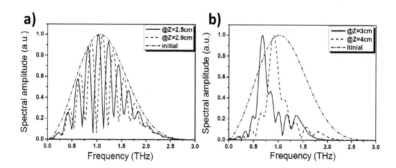

**Figure 14.** (a) Multi-mode theoretical THz spectra obtained with a two circles configuration (radii of 3 mm and 9 mm), at 2.5 cm (solid line) and 2.9 cm (dashed line) from the generating ZnTe. (b) Narrow-band THz theoretical spectra obtained with a five circles configuration (radii of 11, 12, 13, 14 and 15 mm) at 3 cm (solid line) and 4 cm (dashed line) from the generating ZnTe. Each time the initial THz spectrum is shown (dash-dotted line) (taken from [7]).

with 20 mm-large and 30 mm-large ZnTe crystals respectively. Furthermore, due to the initial Gaussian spatial distribution of the pump beam impinging the SLM, NIR circles in the ZnTe crystal have not the same intensity, as a consequence, they generate THz pulses with unequal amplitude. The good compromises in the present experiment were achieved in Fig. 12 when the circle separation was in between 0.5 mm and 2 mm, and in between 0.5 mm and 1 mm in Fig. 13.

## 5. Conclusion

We have presented two new approaches to synthesize THz pulse sequences and narrow-band tunable THz pulses via optical rectification of shaped NIR pulses in ZnTe. In the first one, the NIR pulses are shaped through a simple sinusoidal and/or triangular spectral phase modulation. Our experimental results show a tunability spanning the bandwidth permitted by the phase matching condition in ZnTe, *i.e.* ranging from 0.5 to 2.5 THz, and the possibility to generate spectra with bandwidth as narrow as 140 GHz. These results are in good agreement with a theoretical model taking into account THz generation by optical rectification, the shape of the optical pulses, dispersion in the THz range and the electro-optic detection. The second one has been less explored. It relies on the spatial shaping of the NIR pulse beam profile. Here again, we have demonstrated our ability to generate tunable THz pulses that agreed with our numerical simulations. One should note that this second technique is potentially as versatile as the first one, even if it is not relying on the same kind of pulse shaping method.

One important point is related to the intensity of the THz pulses we can generate using these techniques. A critical issue is the peak intensity that the used crystal can stand. Indeed, since we are dealing with second order nonlinear effects, one could want to increase as much as possible the peak intensity of the NIR femtosecond pulses. Unfortunately, as one increases the peak intensity new nonlinear phenomena are usually taking place. For example, we have recently shown that increasing the peak intensity of the pump beam in ZnTe induces two-photon absorption that impacts quite strongly the optical rectification efficiency as well as the available THz bandwidth [38]. Hence there is a compromise to find between peak power, crystal thickness, THz conversion efficiency. Such an effect becomes negligible if one uses for example LiNbO$_3$ crystal, thanks to its higher optical gap. However, this latter crystal has a lower nonlinear $\chi^{(2)}$, a higher intrinsic THz absorption above 3 THz and a lower THz spectral acceptance. But the latter limitation can be circumvented using the so-called tilted pulsefront phase matching technique [14], leading to THz fields with an amplitude up to 400 kV · cm$^{-1}$ [24]. The combination of tunable THz pulse generation and nonlinear optics makes it possible to envision new developments in THz remote sensing, hyperspectral imaging, study and characterization of THz materials as well as high resolution spectroscopy largely emphasized by recent review articles [13, 14, 34].

## Acknowledgments:

The Conseil Régional d'Aquitaine and the French National Research Agency (ANR 09-BLAN-0212) are acknowledged for financial supports.

## Author details

J. Degert, M. Tondusson, J. Oberlé and É. Freysz
*Université de Bordeaux, Laboratoire Ondes et Matiï£¡re d'Aquitaine, France*

S. Vidal
*Université de Bordeaux, Laboratoire Ondes et Matiï£¡re d'Aquitaine, France*
*Present address: Université de Bordeaux, Centre Lasers Intenses et Applications, France*

C. D'Amico
*Université de Bordeaux, Laboratoire Ondes et Matiï£¡re d'Aquitaine, France*
*Present address: Université Jean Monet, Laboratoire Hubert Curien, France*

## 6. References

[1] Ahn, J.; Efimov, A. V.; Averitt, R. D. & Taylor, A. J. (2003). Terahertz waveform synthesis via optical rectification of shaped ultrafast laser pulses. *Opt. Express*, Vol. 11, 2486-2496, ISSN 1094-4087

[2] Akhmanov, S. A.; Vysloukh, V. A.; Chirkin, A. S. *Optics of Femtosecond Laser Pulses*, AIP, ISBN 0883188511, New York

[3] Amand, T.; Blanchet, V.; Girard, B. & Marie, X. (2005). Coherent Control in Atoms, Molecules and Solids, In: *Femtosecond Laser Pulses*, C. Rullière, (Ed.), 333-394, Springer-Verlag, ISBN 1441918507, New-York

[4] Auston, D. H. ; Cheung, K. P.; Valdmanis, J. A. & D. A. Kleinman, D. A. (1984). Cherenkov Radiation from Femtosecond Optical Pulses in Electro-Optic Media. *Phys. Rev. Lett.*, Vol. 53, 1555-1558, ISSN 0031-9007

[5] Brener, I.; Planken, P. C. M.; Nuss, M. C.; Pfeiffer, L.; Leaird, D. E. & Weiner, A. M. (1993). Repetitive excitation of charge oscillations in semiconductor heterostructures. *Appl. Phys. Lett.*, Vol. 63, 2213-2215, ISSN 0003-6951

[6]   Caumes, J.-P.; Videau, L.; Rouyer, C. & Freysz, E. (2002). Kerr-Like Nonlinearity Induced via Terahertz Generation and the Electro-Optical Effect in Zinc Blende Crystals. *Phys. Rev. Lett.*, Vol. 89, 047401, ISSN 0031-9007

[7]   D'Amico, C.; Tondusson, M.; Degert, J. & Freysz, E. (2009). Tuning and focusing THz pulses by shaping the pump laser beam profile in a nonlinear crystal. *Opt. Express*, Vol. 17, 592-597, ISSN 1094-4087

[8]   Danielson, J. R.; Jameson, A. D.; Tomaino, J. L.; Hui, H.; Wetzel, J. D.; Lee, Y.-S. & Vodopyanov, K. L. (2008). Intense narrow band terahertz generation via type-II difference-frequency generation in ZnTe using chirped optical pulses. *J. Appl. Phys.*, Vol. 104, 03311, ISSN 0021-8979

[9]   Feurer, T.; Vaughan, J. C. & Nelson, K. A. (2003). Spatiotemporal Coherent Control of Lattice Vibrational Waves. *Science*, Vol. 299, 374-377, ISSN 1095-9203

[10]  Feurer, T.; Vaughan, J. C.; Hornung, T. & Nelson, K. A. (2004). Typesetting of terahertz waveforms. *Opt. Lett.*, Vol. 29, 1802-1804, ISSN 0146-9592

[11]  Gallot, G. & Grischkowsky, D. (1999). Electro-optic detection of terahertz radiation. *J. Opt. Soc. Am. B*, Vol. 16, 1204-1212, ISSN 0740-3224

[12]  Gallot, G.; Zhang, J.; McGowan, R. W.; Jeon, T. & Grischkowsky, D. (1999). Measurements of the THz absorption and dispersion of ZnTe and their relevance to the electro-optic detection of THz radiation. *Appl. Phys. Lett.*, Vol. 74, 3450-3452, ISSN 0003-6951

[13]  Hebling, J. ; Yeh, K.-L.; Hoffmann, M. C. & Nelson, K. A. (2008). High Power THz Generation, THz Nonlinear Optics, and THz Nonlinear Spectroscopy. *IEEE J. Selected Topics in Quantum Electronics*, Vol. 14, 345-353, ISSN 1077-260X

[14]  Hoffmann, M.C. & Fülöp, J.A. (2001). Intense ultrashort terahertz pulses: Generation and applications. *J. Phys. D: Appl. Phys.*, Vol. 44, 083001, ISSN 0022-3727

[15]  Huggard, P. G.; Cluff, J. A.; Shaw, C. J.; Andrews, S. R.; Linfield, E. H. & Ritchie, D. A. (1997). Coherent control of cyclotron emission from a semiconductor using sub-picosecond electric field transients. *Appl. Phys. Lett.*, Vol. 71, 2647-2649, ISSN 0003-6951

[16]  Hurlbut, W. C.; Norton, B. J.; Amer, N. & Lee, Y.-S. (2006). Manipulation of terahertz waveforms in nonlinear optical crystals by shaped optical pulses. *J. Opt. Soc. B*, Vol. 23, 90-93, ISSN 0740-3224

[17]  Kohli, K. K.; Vaupel, A.; Chatterjee, S. & Rühle, W. W. (2009). Adaptive shaping of THz-pulses generated in $\langle 100 \rangle$ ZnTe crystals. *J. Opt. Soc. B*, Vol. 26, 74-78, ISSN 0740-3224

[18]  Lee, Y.-S.; Meade, T.; Norris, T. B. & Galvanauskas, A. (2001). Tunable narrow-band terahertz generation from periodically poled lithium niobate. *Appl. Phys. Lett.*, Vol. 78, 3583-3585, ISSN 0003-6951

[19]  Lee, Y.-S.; Amer, N. & Hurlbut, W. C. (2003). Terahertz pulse shaping via optical rectification in poled lithium niobate. *Appl. Phys. Lett.*, Vol. 82, 170-172, ISSN 0003-6951

[20]  Lee, Y.-S. (2009). *Principles of Terahertz Science and Technology*, Springer-Verlag, ISBN 038709539X, New-York.

[21]  Liu, Y.; Park, S.-G. & Weiner, A. M. (1996). Terahertz Waveform Synthesis via Optical Pulse Shaping. *IEEE J. Sel. Top. Quantum Electron.*, Vol. 2, 709-719, ISSN 1077-260X

[22]  Monmayrant, A. & Chatel, B. (2004). New phase and amplitude high resolution pulse shaper. *Rev. Sci. Instrum.*, Vol. 75, 2668-2671, ISSN 0034-6748

[23] Monmayrant, A.; Weber, S. & Chatel, B. (2010). A newcomer's guide to ultrashort pulse shaping and characterization. *J. Phys. B: At. Mol. Opt. Phys.*, Vol. 43, 103001, ISSN 0953-4075

[24] Nagai, M.; Matsubara, E. & Ashida, M. (2012). High-efficiency terahertz pulse generation via optical rectification by suppressing stimulated Raman scattering process. *Opt. Express*, Vol. 20, 6509-6514, ISSN 1094-4087

[25] Nahata, A.; Weling, A. S. & Heinz, T. F. (1996). A wideband coherent terahertz spectroscopy system using optical rectification and electro-optic sampling. *Appl. Phys. Lett.*, Vol. 69, 2321-2323, ISSN 0003-6951

[26] Planken, P. C. M.; Brener, I.; Nuss, M. C.; Luo, M. S. C. & Chuang, S. L. (1993). Coherent control of terahertz charge oscillations in a coupled quantum well using phase-locked optical pulses. *Phys. Rev. B*, Vol. 48, 4903-4906, ISSN 1098-0121

[27] Qi, T.; Shin, Y.-H.; Yeh, K.-L.; Nelson, K. A. & Rappe, A. M. (2009). Collective Coherent Control: Synchronization of Polarization in Ferroelectric $PbTiO_3$ by Shaped THz Fields. *Phys. Rev. Lett.*, Vol. 102, 247603, ISSN 0031-9007

[28] Reimann, K. (2007). Table-top sources of ultrashort THz pulses. *Rep. Prog. Phys.*, Vol. 70, 1597-1632, ISSN 0034-4885.

[29] Renard, M.; Chaux, R.; Lavorel, B. & Faucher, O. (2004). Pulse trains produced by phase-modulation of ultrashort optical pulses: tailoring and characterization. *Opt. Express*, Vol. 12, 473-482, ISSN 1094-4087

[30] Sohn, J. Y.; Ahn, Y. H.; Park, D. J.; Oh, E. & Kim, D. S. (2002). Tunable terahertz generation using femtosecond pulse shaping. *Appl. Phys. Lett.*, Vol. 81, 13-15, ISSN 0003-6951

[31] Stepanov, A. G.; Hebling, J. & Kuhl, J. (2004). Generation, tuning, and shaping of narrow-band, picosecond THz pulses by two-beam excitation. *Opt. Express*, Vol. 12, 4650-4658, ISSN 1094-4087

[32] Stepanov, A. G.; Hebling, J. & Kuhl, J. (2005). THz generation via optical rectification with ultrashort laser pulse focused to a line. *Appl. Phys. B*, Vol. 81, 23-26, ISSN 0946-2171

[33] Stobrawa, G.; Hacker, M.; Feurer, T.; Zeidler, D.; Motzkus, M. & Reichel, F. (2001). A new high-resolution femtosecond pulse shaper. *Appl. Phys. B*, Vol. 72, 627-630, ISSN 0946-2171

[34] Tanaka, M.; Hirori, H. & Nagai, M. (2011). THz Nonlinear Spectroscopy of Solids. *IEEE Transactions on Terahertz Science and Technology*, Vol. 1, 301-312, ISSN 2156-342X

[35] Vaughan, J. C.; Feurer, T.; Stone, K. W. & Nelson, K. A. (2006). Analysis of replica pulses in femtosecond pulse shaping with pixelated devices. *Opt. Express*, Vol. 14, 1314-1328, ISSN 1094-4087

[36] Vaughan, J. C.; Hornung, T.; Feurer, T. & Nelson, K. A. (2005). Diffraction-based femtosecond pulse shaping with a two-dimensional spatial light modulator. *Opt. Lett.*, Vol. 30, 323-325, ISSN 0146-9592

[37] Vidal, S.; Degert, J.; Oberli£¡, J. & Freysz, E. (2010). Femtosecond optical pulse shaping for tunable terahertz pulses generation. *J. Opt. Soc. Am. B.*, Vol. 27, 1044-1050, ISSN 0740-3224

[38] Vidal, S.; Degert, J.; Tondusson, M.; Oberli£¡, J. & Freysz, E. (2011). Impact of dispersion, free carriers and two-photon absorption on the generation of intense THz pulses in ZnTe crystals. *Appl. Phys. Lett.*, Vol. 98, 191103, ISSN 0003-6951

[39] Wahlstrand, J. K. & Merlin, R. (2003). Cherenkov radiation emitted by ultrafast laser pulses and the generation of coherent polaritons. *Phys. Rev. B*, Vol. 68, 054031, ISSN 1098-0121

[40] Ward, D. W.; Beers, J. D.; Feurer, T.; Statz, E. R.; Stoyanov, N. S. & Nelson, K. A. (2004). Coherent control of phonon-polaritons in a terahertz resonator fabricated with femtosecond laser machining. *Opt. Lett.*, Vol. 29, 2671-2673, ISSN 0146-9592

[41] Weiner, A. M.; Leaird, D. E.; Wiederrecht, G. P. & Nelson, K. A. (1990). Femtosecond Pulse Sequences Used for Optical Manipulation of Molecular Motion. *Science*, Vol. 247, 1317-1319, ISSN 1095-9203

[42] Weiner, A. M. (2000). Femtosecond pulse shaping using spatial light modulators. *Rev. Sci. Instrum.*, Vol. 71, 1929-1960, ISSN 0034-6748

[43] Yamaguchi, M. & Das, J. (2009). Terahertz wave generation in nitrogen gas using shaped optical pulses. *J. Opt. Soc. B*, Vol. 26, 90-94, ISSN 0740-3224

[44] Yano, R.; Nakagawa, K. & Shinojima, H. (2009). Phase and Amplitude Control of Free Induction Decay Emitted from Water Vapor at 0.55 THz Transition. *Jpn. J. Appl. Phys.*, Vol. 48, 022401, ISSN 0021-4922

# Permissions

The contributors of this book come from diverse backgrounds, making this book a truly international effort. This book will bring forth new frontiers with its revolutionizing research information and detailed analysis of the nascent developments around the world.

We would like to thank Igor Peshko, for lending his expertise to make the book truly unique. He has played a crucial role in the development of this book. Without his invaluable contribution this book wouldn't have been possible. He has made vital efforts to compile up to date information on the varied aspects of this subject to make this book a valuable addition to the collection of many professionals and students.

This book was conceptualized with the vision of imparting up-to-date information and advanced data in this field. To ensure the same, a matchless editorial board was set up. Every individual on the board went through rigorous rounds of assessment to prove their worth. After which they invested a large part of their time researching and compiling the most relevant data for our readers. Conferences and sessions were held from time to time between the editorial board and the contributing authors to present the data in the most comprehensible form. The editorial team has worked tirelessly to provide valuable and valid information to help people across the globe.

Every chapter published in this book has been scrutinized by our experts. Their significance has been extensively debated. The topics covered herein carry significant findings which will fuel the growth of the discipline. They may even be implemented as practical applications or may be referred to as a beginning point for another development. Chapters in this book were first published by InTech; hereby published with permission under the Creative Commons Attribution License or equivalent.

The editorial board has been involved in producing this book since its inception. They have spent rigorous hours researching and exploring the diverse topics which have resulted in the successful publishing of this book. They have passed on their knowledge of decades through this book. To expedite this challenging task, the publisher supported the team at every step. A small team of assistant editors was also appointed to further simplify the editing procedure and attain best results for the readers.

Our editorial team has been hand-picked from every corner of the world. Their multi-ethnicity adds dynamic inputs to the discussions which result in innovative outcomes. These outcomes are then further discussed with the researchers and contributors who give their valuable feedback and opinion regarding the same. The feedback is then collaborated with the researches and they are edited in a comprehensive manner to aid the understanding of the subject.

Apart from the editorial board, the designing team has also invested a significant amount of their time in understanding the subject and creating the most relevant covers. They scrutinized every image to scout for the most suitable representation of the subject and create an appropriate cover for the book.

The publishing team has been involved in this book since its early stages. They were actively engaged in every process, be it collecting the data, connecting with the contributors or procuring relevant information. The team has been an ardent support to the editorial, designing and production team. Their endless efforts to recruit the best for this project, has resulted in the accomplishment of this book. They are a veteran in the field of academics and their pool of knowledge is as vast as their experience in printing. Their expertise and guidance has proved useful at every step. Their uncompromising quality standards have made this book an exceptional effort. Their encouragement from time to time has been an inspiration for everyone.

The publisher and the editorial board hope that this book will prove to be a valuable piece of knowledge for researchers, students, practitioners and scholars across the globe.

# List of Contributors

**Akira Endo**
Research Institute for Science and Engineering, Waseda University, Tokyo, Japan
HiLASE Project, Institute of Physics AS CR, Prague, Czech Republic

**Emmanuel d'Humières**
Université de Bordeaux – CEA - CNRS - CELIA, France

**Evgenii Gorokhov, Kseniya Astankova, Alexander Komonov and Arseniy Kuznetsov**
Institute of Semiconductor Physics of SB RAS, Russia
Laser Zentrum Hannover, Germany

**V. V. Apollonov**
Prokhorov General Physics Institute of RAS, Moscow, Russia

**Kun Huang, E Wu, Xiaorong Gu, Haifeng Pan and Heping Zeng**
State Key Laboratory of Precision Spectroscopy, East China Normal University, China

**Guofeng Zhang, Ruiyun Chen, Yan Gao, Liantuan Xiao and Suotang Jia**
State Key Laboratory of Quantum Optics and Quantum Optics Devices, Laser Spectroscopy Laboratory, Shanxi University, Taiyuan, China

**J. Degert, M. Tondusson, J. Oberlé and É. Freysz**
Université de Bordeaux, Laboratoire Ondes et Matiï£¡re d'Aquitaine, France

**S. Vidal**
Université de Bordeaux, Laboratoire Ondes et Matiï£¡re d'Aquitaine, France
Present address: Université de Bordeaux, Centre Lasers Intenses et Applications, France

**C. D'Amico**
Université de Bordeaux, Laboratoire Ondes et Matiï£¡re d'Aquitaine, France
Present address: Université Jean Monet, Laboratoire Hubert Curien, France

Printed in the USA
CPSIA information can be obtained
at www.ICGtesting.com
JSHW011811301024
72690JS00002B/51

9 781632 400147